THEMES OF THE Times

ON THE

Environment

VOLUME 2

A COLLECTION OF ARTICLES FROM

The New York Times

San Francisco Boston New York
Capetown Hong Kong London Madrid Mexico City
Montreal Munich Paris Singapore Sydney Tokyo Toronto

Editor-in-Chief: Beth Wilbur
Senior Acquisitions Editor: Chalon Bridges
Project Editor: Nora Lally-Graves
Director of Development: Deborah Gale
Managing Editor: Michael Early
Production Supervisor: Lori Newman
Production Management and Composition: Carlisle Publishing Service
Copyeditor: Sue Grutz
Manufacturing Buyer: Stacy Wong
Executive Marketing Manager: Lauren Harp
Text and Cover Printer: Courier Stoughton

All pedagogical content, including article summaries, *In Review* article-related questions, and answers to *Testing Your Comprehension* and *Interpreting Graphs and Data* questions, was created by Kristy Manning.

The paper used for this publication is recycled, containing up to 20% post-consumer waste.

ISBN 0-8053-9608-X

All articles in the *Themes of the Times on the Environment* are copyrighted by The New York Times Company and reprinted with permission.

All pedagogical content for this supplement has been created by Pearson Benjamin Cummings, and includes: article summaries and cross-references to *Environment: The Science behind the Stories*, Second Edition, and *Essential Environment: The Science behind the Stories*, Second Edition, in the Table of Contents; topics grid referencing subjects covered in the articles; "*In Review*" questions that follow each; and the "Answers to *Testing Your Comprehension Questions*" and "*Interpreting Graphs and Data*."

Copyright © 2007 Pearson Education, Inc., publishing as Pearson Benjamin Cummings, 1301 Sansome St., San Francisco, CA 94111. All rights reserved. Manufactured in the United States of America. This publication is protected by Copyright and permission should be obtained from the publisher prior to any prohibited reproduction, storage in a retrieval system, or transmission in any form or by any means, electronic, mechanical, photocopying, recording, or likewise. To obtain permission(s) to use material from this work, please submit a written request to Pearson Education, Inc., Permissions Department, 1900 E. Lake Ave., Glenview, IL 60025. For information regarding permissions, call (847) 486-2635.

Many of the designations used by manufacturers and sellers to distinguish their products are claimed as trademarks. Where those designations appear in this book, and the publisher was aware of a trademark claim, the designations have been printed in initial caps or all caps.

Pearson Benjamin Cummings is a trademark, in the U.S. and/or other countries, of Pearson Education, Inc. or its affiliates.

www.aw-bc.com

1 2 3 4 5 6 7 8 9 10—CRS—10 09 08 07 06

THEMES OF THE Times ON THE Environment

Contents

THEMES OF THE Times ON THE Environment

	Conservation	Biotechnology	Energy	Global Warming	Pollution	Population	Resources	Sustainable Solutions
Clogged Rockies Highway Divides Coloradans, p. 1						✓		
Who Will Work the Farms? Immigrants Wanted: Legal Would Be Nice, but Illegal Will Suffice, p. 5						✓		
Biotech's Sparse Harvest; A Gap Between the Lab And the Dining Table, p. 9		✓						
Deals Turn Swaths of Timber Company Land Into Development-Free Areas, p. 13	✓						✓	
Americans Are Cautiously Open to Gas Tax Rise, Poll Shows, p. 16			✓	✓	✓			
The New Face Of an Oil Giant; Exxon Mobil Style Shifts a Bit, p. 19			✓	✓	✓			
Corn Farmers Smile as Ethanol Prices Rise, but Experts on Food Supplies Worry, p. 23			✓				✓	✓
Gualeguaychú Journal; A Back-Fence Dispute Crosses an International Border, p. 26					✓			
Recycled Inspection Reports and Other Water System Irregularities Stir Concerns Upstate, p. 29							✓	
Eating Well: Advisories on Fish and the Pitfalls of Good Intent, p. 32					✓		✓	
Out of Old Mines' Muck Rises New Reclamation Model for West, p. 36							✓	✓
New Uses for Glut of Small Logs From Thinning of Forests, p. 39	✓						✓	✓
Canada to Shield 5 Million Forest Acres, p. 42	✓						✓	✓
Competing Plans to Repair New Orleans Flood Protection, p. 45								✓

Clogged Rockies Highway Divides Coloradans

By Kirk Johnson
***The New York Times,* January 25, 2006**

Correction Appended

SILVERTHORNE, Colo., Jan. 22—When Interstate 70 was built through here in the 1960s and 70's, the Colorado Rockies were largely rural and remote, and the old roads that the highway replaced were a widely recognized danger.

"Do we accommodate growth, or do we stifle it?"

Over the years, as the population grew, delays and frustrations on the highway began to mount. Traffic jams at nosebleed altitude became common. In 8,800-foot-high Silverthorne, which was little more than a gas station pit stop a generation ago, with a grocery that got fresh produce only on Thursdays, alpine meadows gave way to factory outlet stores.

Now state officials are considering a major and contentious widening project for Interstate 70 that is dividing people over the question of who the highway is for and how it transformed these mountains.

The project is a variation of a drama that is playing out across much of the West as once-rural outposts are transformed into brimming settlements with newfound political and economic clout in transportation decisions.

As the Federal Highway Act of 1956 established the Interstate System and helped open vast expanses of the West, highways like Interstate 70 changed just about everything by putting on the map distant places that had been mostly untouched.

Photographs by Carmel Zucker for The New York Times

Traffic snaked past Idaho Springs, Colo., on Interstate 70 recently, heading east out of the mountains.

Now, the very places that were changed, like Silverthorne—65 miles west of Denver—are wading in as aggressive and muscular participants in discussions about what comes next.

In Nevada, the expansion U.S. 95 connecting Las Vegas and its sprawling western suburbs has resumed after environmentalists settled a lawsuit last year over the effect of increased

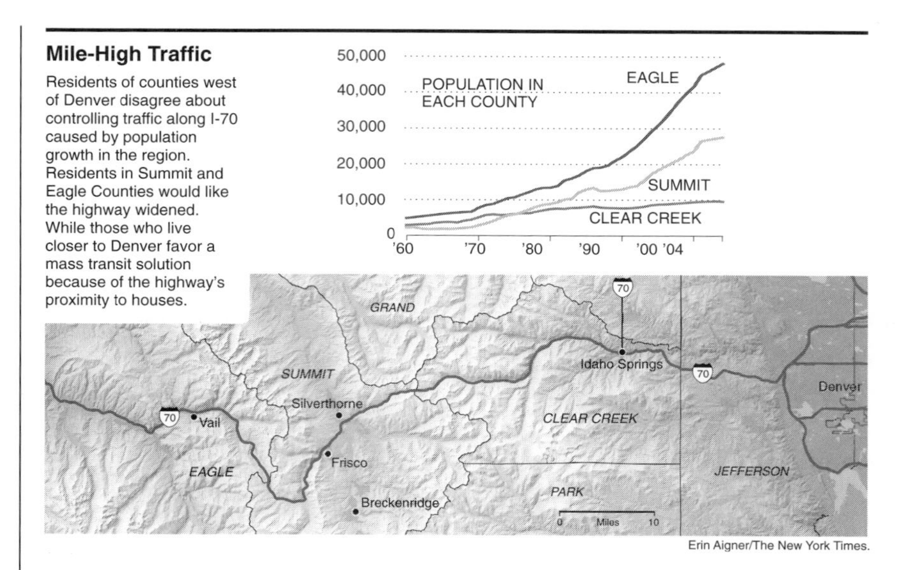

Erin Aigner/The New York Times.

Bill and Linda Wellington plan to move to northeastern Colorado to avoid Summit County congestion.

vehicle emissions on nearby residents.

The proposed Legacy Highway in Utah, extending south of Salt Lake City, was blocked for years before receiving final approval this month.

Elsewhere in Colorado, a plan for a toll road across the once-empty plains east of Denver was put on hold last year after opposition from residents.

Transportation experts, politicians and residents agree that the stakes and implications of these fights are enormous, touching on tenets of the West that are scriptural: unbridled growth, local identity, civic autonomy and an uneasy dependence on government.

Here in Colorado, where Interstate 70 is crucial to the tourism economy and the state's image as the mountain playground of the West, businesses, residents and interest groups do not remotely agree on what to do.

"The question is, How much do we really want to improve I-70—and do we want to improve it so much that it changes the character of our communities?" said Gary Severson, the executive director of the Northwest Colorado Council of Governments. "That's the tightrope."

The idea that transportation systems can reshape the regions through which they pass is well established. As far back as the Erie Canal and the transcontinental railroad, geographers and historians say that getting from Point A to Point B has always been at least partly about property values, boosterism and the restless American impulse to move on and create anew.

But that pattern is being given a decidedly new twist here on a road that was hailed and agonized over as one of the most daunting stretches of the Interstate System.

Places like Silverthorne—incorporated in 1967, when it was populated mainly by highway construction workers—have become destinations for shopping or homebuilding and tend to favor an expansion of the highway that would deliver more of the bounty that Interstate 70 has already bestowed. Towns closer to Denver, like Idaho Springs, which have not seen the influx of vacation homes or tourists and which also have many more commuters to Denver, say mass transit must be on the table.

Some politicians and residents say that doing nothing may be the wisest course. Colorado's population in and out of the mountains is expected to increase by 50 percent in the next 20 years, pushing people toward mass transit as traffic congestion worsens.

Not everything that has happened here is a result of transportation, of course. The explosion of population and the economy since the completion of Interstate 70's last leg in 1979—the second tube of the Eisenhower/Johnson Tunnel at 11,000 feet across the Continental Divide—also coincided with a demographic transformation as members of the post-World War II generation reached their peak earning years, stock and real estate markets boomed and changes in tax law made buying vacation real estate more attractive.

Resorts that opened or grew during the highway's early years—like Vail, in 1962, Keystone in 1970 and Beaver Creek in 1980—have moved toward year-round operations and real-estate development that bring more visitors and residents.

The results have all fed back into the equation of traffic, which is increasingly bumper-to-bumper on weekends in winter and summer. Another record was set last year for the number of cars trying to squeeze through the four lanes of the Eisenhower/Johnson Tunnel in one month—just over 37,000 a day, last July.

Issues of class and clout have further clouded the picture.

Summit County, where Silverthorne is located, was one of the fastest-growing counties in the nation in the 1990's, with the population up 83 percent. Equally important, many residents and politicians say, is that the nine counties along the Interstate 70 corridor—led by Summit and Eagle, where the big resorts and the biggest waves of homebuilding are concentrated—are becoming more and more important to the state's economy.

In 2000, a private study commissioned by the state found that the corridor produced about $2 billion in recreation revenue, out of $9.3 billion statewide, and generated $136 million in state and local tax revenues.

"Do we accommodate growth, or do we stifle it? That is the question," said Lou DelPiccolo, Silverthorne's mayor, who favors a highway widening plan.

Thirty miles east, in Clear Creek County, where the growth has been slower and the big payoff from the highway never quite materialized, political leaders like the Idaho Springs mayor, Dennis Lunbery, are asking the opposite question: Could they survive the widening of the highway?

Idaho Springs, confined in a narrow canyon—the back door of City Hall is less than 100 feet from the Interstate's westbound guardrail—would be destroyed, Mr. Lunbery said, by the miasma of asphalt, noise and dust that a bigger highway, and the years of construction, would bring.

Mr. Lunbery said he thought that the state, in its environmental assessment of the corridor, had "stacked the deck" toward a highway-widening option by excluding any plan that costs more than $4 billion. Most mass transit proposals would add at least another $2 billion to $4 billion.

Coincidentally or not, widening is what the destination communities farther west mostly want. Many people farther west also worry that a mass transit rail line, however good it may be for closer-in communities like Idaho Springs, could turn resort communities into bedroom communities, full of commuters catching the train to their offices in Denver.

The executive director of the State Department of Transportation, Thomas E. Norton, who is expected to make a decision on the corridor later this year, said he was open to all options.

"I don't think there is a bias in my perspective," Mr. Norton said. "There is a bias toward best use of the public dollar, and until you can get really high transit kinds of usage, the economics are not there. Rail that can move 10,000 people an hour looks great, but if only 12 days of the year it would be used by that many people, it's not an efficient use of dollars."

Many people, including Mr. Lunbery, who expects the state's study to be challenged in court almost no matter what happens, think that nothing will happen for years on the corridor—partly because of the difficulty in finding the money—and that worsening traffic is, for now, the only certainty.

Gary Lindstrom thinks that is a good thing. He is a legislator in the Colorado House whose district includes much of the corridor. He is also a Democratic candidate for governor this year and favors the mass transit idea, but he agrees with Mr. Norton that for now it does not appear feasible. Time will change that equation, Mr. Lindstrom said.

"The worst thing we can do is widen the highway," he said. "We need to keep the congestion so people will be interested in the transit."

No matter what unfolds, Linda Wellington and her husband, Bill, who have lived all their lives here, will probably not see it. Ms. Wellington has watched it all—the good and the bad—and often talked it over with her father, a career highway worker, before he died, who regularly plowed the highest passes of the old road before Interstate 70 was built.

"At one time I was bitter. I'd say, 'Dad, don't you hate all the people and the traffic?' and he'd say, 'We need to move forward,'" said Ms. Wellington, who is in her 50's. "So that's where I am now. You can't stop it. It's here, we've built it, and you can't close the door."

The Wellingtons are planning to move. Mr. Wellington, who is 56, said he had been looking at a place in northeastern Colorado, out on the plains toward Kansas, where it is not as crowded as the mountains have become and where their son now lives. He will hate to leave, he said, but things have changed too much to stay.

Correction: January 28, 2006, Saturday. An article on Wednesday about a plan to widen Interstate 70 in Colorado misidentified the last section of the highway to be built and the year of its completion. It was a section near Glenwood Springs, in 1992, not the second tube of the Eisenhower/Johnson Memorial Tunnels, in 1979.

Copyright 2006 The New York Times Company. Reprinted with permission.

In Review: Clogged Rockies Highway Divides Coloradans

Testing Your Comprehension

1. The article refers to long-held tenets of the West in referring to the stakes and implications of the growth fights. What are those tenets?
2. What is the "tightrope" that Gary Severson describes as the dilemma facing Colorado's planners?
3. What are the three things that characterize movement and transportation in this country as far back as the Erie Canal and the transcontinental railroad?

4. In general, what is the difference between towns closer to Denver versus those farther away in terms of their attitude toward the highway expansion?
5. By how much is Colorado's population in and out of the mountains expected to increase over the next 20 years?
6. Aside from transportation, what other factors coincided with the area's explosion of population and the economy over the past 25 years?
7. According to a private study conducted in 2000, how much money did the I-70 corridor produce in recreation and tax revenues?
8. Some communities are concerned that they would not survive the widening of the highway. Why?
9. In addition to the economic benefits that a widened highway could bring, why are destination communities farther west more in favor of widening than mass transit?
10. What is the concern of Thomas Norton, the executive director of the State Department of Transportation, about jumping into mass transit prematurely?

Weighing the Issues

1. List the pros and cons of widening the highway versus pursuing mass transit.
2. Geographers and historians say that "getting from Point A to Point B has always been at least partly about property values, boosterism and the restless American impulse to move on and create anew." Do you think this will ever change, and if so, under what circumstances?
3. According to the article, many people farther west worry that a mass transit line could flood their towns with commuters. Why do you think these towns are opposed to becoming "bedroom communities"?
4. The State Department of Transportation is concerned that not enough people will take mass transit to justify the expense of its operations. What steps could be taken to encourage ridership?
5. Do you think the highway should be widened? Why or why not? If so, under what circumstances?

More to the Story

Map the planned widening of I-70 and the build out of the Denver Metro transit system to get a better picture of the I-70 corridor. Which towns will benefit and which will not? In what ways (e.g., environmentally, economically)? Think about the types of industries that will move into the towns farther west of Denver. To what extent will these industries contribute to the tax base and help prevent the further expansion of "bedroom communities" throughout the corridor? What are the benefits and drawbacks of these communities, particularly in terms of their economy? Do you think mass transit is a viable option for mitigating against further traffic? Why or why not?

Interpreting Graphs and Data

1. Which of the three counties referenced in the graph has the fastest rate of growth?
2. Which county has the slowest rate of growth?
3. Which county has shown the greatest rate of growth since 2000?
4. How do you expect the widening of I-70 to affect the rates of growth of each of the three counties?

Useful Websites

http://www.nwc.cog.co.us/

http://www.dot.state.co.us/Communications/news/I70PEISFactSheet.pdf

http://www.metrodenver.org/DataCenter/Infrastructure/Transportation/MassTransit.icm

http://maps.google.com/maps?oi=map&q=Silverthorne,+CO

Who Will Work the Farms?

Immigrants Wanted: Legal Would Be Nice, but Illegal Will Suffice

By Eduardo Porter

***The New York Times*, March 23, 2006**

KINSTON, N.C.—In the plainspoken manner common to her fellow farmers, Faylene Whitaker has a message for members of Congress struggling to overhaul the nation's immigration law.

"We would rather use legal workers," said Ms. Whitaker, who grows tobacco, tomatoes and other crops on the 500-acre farm she and her husband own in the Piedmont region of North Carolina. But "if we don't get a reasonable guest worker program we are going to hire illegals."

Ms. Whitaker knows what she is talking about. Indeed, she is one of the few farmers in the United States to employ legal immigrant guest workers. Others are not exactly lining up to join the program.

The experience of farmers like Ms. Whitaker underscores why most experts say that simply tightening the border with Mexico will not work.

To curb illegal immigration, they say, the government needs both carrots and sticks: legal work channels for future immigrants and the current population of workers unlawfully in the country along with a much more effective bar against hiring illegal labor.

"The punch line is, guest workers are going to cost you more money," said Phil Martin, professor of agricultural and resource economics at the University of California, Davis. For a guest worker program to work, he added, "the prerequisite is you have to have illegal immigration under control."

When immigration rules were last overhauled in 1986—legalizing more than three million unlawful immigrants and supposedly banning the employment of illegal workers in the future—Congress allowed for a vast seasonal guest worker plan. It was intended to ensure that farmers could continue to get the cheaper foreign labor they wanted, but in a legally acceptable way.

It has not worked out. Despite expectations that growers would flock to the guest worker program, named H-2A, it never took off: fewer than 25,000 farm workers came to the United States on temporary guest worker visas in 2004, the last year with figures available.

It is not as if growers do not need the foreign labor: according to farmers' own estimates, about 70 percent of the 1.2 million hired workers tilling fields and picking crops are illegal immigrants.

What went wrong? Despite the ostensible sanctions against businesses that hire illegal workers, few employers were ever punished. So rather than go through the more complex process of importing temporary workers legally, most chose the cheaper and generally

Photographs by Karen Tam for The New York Times

A Mexican flag hangs in Duplin Country, N.C., Arturo Martinex Mendoza, left, is an illegal immigrant, while José Aharón Rivera Medrano has "guest" status.

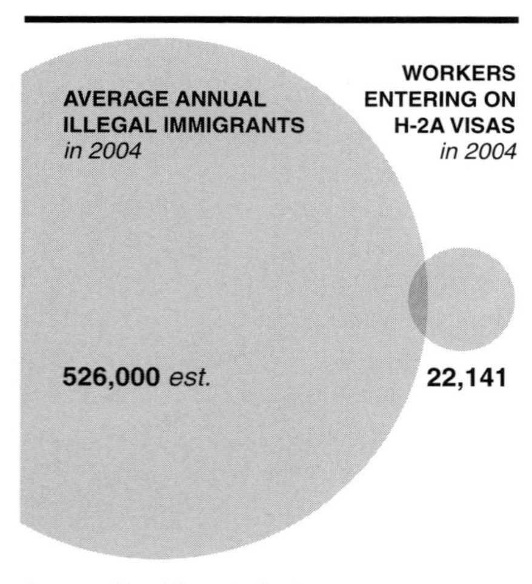

Sources: Pew Hispanic Center, U.S. Citizenship and Immigration Services

The New York Times

risk-free route of hiring illegal migrants who presented themselves with forged papers.

Congress is once again debating how to cut off illegal immigration but this time it has to deal with as many as 11 million unlawful immigrants in the United States.

Some lawmakers are talking about erecting a high wall along the entire 2,000-mile border separating the United States from Mexico. Others are pushing to declare all immigrants here illegally felons who would never be allowed to take up legal residence in the United States.

But as long as the government looks the other way while employers are allowed to fill their demands for cheap labor from a deep pool of foreigners desperate to find work, hundreds of thousands more are expected to manage one way or another to cross the border illegally every year in search of jobs.

"If there is a need for workers, they will find a way in, if you don't have a method for satisfying the legitimate needs of American employers," said Senator Arlen Specter, Republican from Pennsylvania and chairman of the Judiciary Committee, which has been trying to hammer out a new immigration bill to present to the full Senate as early as next Monday.

But this time, Mr. Specter added, any new law must contain tougher sanctions against an employer who "knowingly hires someone that is not here legally."

Like most employers of cheap labor, North Carolina's growers say they need immigrant workers to stay afloat. "We need a guest worker program because somebody's got to work," said Kendall Hill, who grows tobacco, sweet potatoes and other crops on 4,000 acres here in Kinston.

He gazes across the flat dirt of his farm toward the sheds that will house up to 120 immigrant H-2A workers at the peak of the sweet potato harvest in September. "If not for Mexican workers," he adds, "this country would be in chaos."

As for illegal immigrants, many say that they would welcome a plan that allowed them to work legally in the United States and return to their families in Mexico every year.

Arturo Martínez Mendoza, an illegal immigrant harvesting mustard for $1.25 a box on a farm near Faison, said he had even tried to get into the program. "It was full," he said, "so I came in as a wetback."

José Aharón Rivera Medrano, an H-2A worker from Mexico's northern state of Durango, has worked on Mr. Hill's farm 10 months each year for the last 11 years. He is happy about the higher wages of $500 to $600 a week he receives under the guest worker program than if he were working illegally, but the most important thing, he said, "is that you don't risk your life crossing the border."

. . . fewer than 25,000 farm workers came to the United States on temporary guest worker visas in 2004

According to Leticia Zavala, southern organizing director for the union known as the Farm Labor Organizing Committee, there is a waiting list of 17,000 Mexican workers who have come to North Carolina in the past and would like to return.

But the guest worker program has failed to bridge the gap between the farmers and the workers. Farmers complain that the program is simply too expensive. It sets a floor for pay—a regional average of the wages in several farm occupations—that will rise this year to $8.51 an hour from $8.24 in 2005. Farmers must also house the workers and pay workers' compensation.

Moreover, after a lawsuit spearheaded by the union against North Carolina growers, a state court ordered farmers to pay all recruitment costs, including the H-2A visa, recruiters' fees and round-trip transportation from home—adding up to some $900 a worker.

"I've been 10 years on the program but I'm reaching the point of diminishing returns," said Sam Crews, who employs six guest workers on his North Carolina tobacco farm near the border with Virginia.

The guest worker plan has some advantages for growers: legal workers are more dependable than illegal ones, they say. And it relieves them of any lingering fears that their harvest might be in jeopardy if immigration authorities conduct raids that disrupt the supply of illegal workers.

Yet growers have so many other complaints about the program that many are simply dropping out. To be entitled to use guest workers, they must certify to the Labor Department that they cannot find American workers for the job, a process that they say is overly onerous.

And once non-H-2A immigrant workers are brought out of the shadows, farmers who employ legal guest workers often find themselves being sued over working conditions by immigrants' rights groups and other foes of the program.

Bruce Goldstein, executive director of the Farm Workers Justice Fund in Washington, which has brought lawsuits against farmers, says there are good reasons for such lawsuits.

"We need a guest worker program because somebody's got to work," said Kendall Hill, who farms 4,000 acres in Kinston, N.C.

"Farmers under the H-2A program are being sued when they violate the law," Mr. Goldstein said. "They frequently violate the law. And employers have tremendous bargaining power over workers who are too fearful to challenge unfair or illegal conduct."

For many farmers, then, illegal immigrants simply provide a cheaper alternative that involves far less bother than the guest worker program. North Carolina farmers can harvest their crops paying $6 to $6.50 an hour—no workers comp, no recruitment fees—to "green card" workers or "otherwise documented" workers. Those are euphemisms for immigrants unlawfully in the United States, who typically show up with a fake green card to get a job.

"Workers provide us with ID and Social Security cards to keep me in compliance and keep them in compliance, but I have no clue of their provenance," said one Wilson County tobacco, cotton and sweet potato farmer who spoke only on condition of anonymity to avoid exposing his work force to scrutiny from the authorities.

Farmers in North Carolina still employ more legal guest workers than in any other state, but their numbers are dwindling. At their peak in 2000, over 1,000 farmers in the growers' association employed 10,000 guest workers. This year, about 500 farmers are left, and they will import only about 5,000 workers.

Some of those missing farmers were driven out by increased competition in the global tobacco market. Others have simply switched to illegal immigrant workers. And if things do not change, many farmers say, the system may well disappear altogether.

"When the H-2A program prices itself out, I'm not growing anything else," said Lee Whicker, who employs a dozen guest workers on his tobacco farm near Fayetteville. "I'm done with it."

Copyright 2006 The New York Times Company. Reprinted with permission.

In Review: Who Will Work the Farms? *Immigrants Wanted: Legal Would Be Nice, But Illegal Will Suffice*

Testing Your Comprehension

1. Ms. Whitaker and other farmers believe that to curb illegal immigration, the government needs both "carrots" and "sticks." What do they mean by "carrots" and "sticks"?
2. According to Phil Martin, what is the primary problem for farmers participating in the guest worker program, and what needs to occur for a guest worker program to work?
3. In 1986, Congress allowed for a vast seasonal guest worker plan. What was its goal?
4. How many farm workers participated in the program in 2004?
5. According to farmers' own estimates, how many hired workers tilling fields and picking crops are illegal immigrants?
6. Why has the guest worker program been ineffective?
7. List four aspects of the guest worker program that contribute to its cost.
8. What are the advantages of the guest worker plan to growers?
9. Describe two other complaints that growers have about the program.
10. In the state of North Carolina, what has happened to the numbers of farmers employing legal guest workers, and the number of guest workers, since 2000?

Weighing the Issues

1. Faylene Whitaker is one of the few farmers in the United States to employ legal immigrant guest workers. Why do you think she has chosen to participate in the program when others do not?
2. According to Kendall Hill of Kinston, "If not for Mexican workers . . . . this country would be in chaos." Do you agree? Why or why not?
3. What environmental impacts, if any, are caused by illegal immigration?
4. The article describes several options to manage illegal immigration in the United States. If you were a member of Congress, what would you recommend?
5. How could the guest worker program be improved to make it easier for employers to participate?

More to the Story

Develop a brief critique of the guest worker program covered by CNN.com on March 31, 2006. How do you think this will affect the numbers of illegal immigrants entering the United States each year? What will be the impact on employment in this country? How will the program affect the environment? What will this do to rural communities like Faylene Whitaker's? If you were Faylene Whitaker, would you participate in the program? Why or why not?

Useful Websites

http://www.whitehouse.gov/news/releases/2004/01/20040107-3.html

http://www.doleta.gov/business/gw/guestwkr/

http://www.washingtonpost.com/wp-dyn/content/article/2005/10/18/AR2005101801613.html

http://www.cnn.com/2006/POLITICS/03/31/bush.cancun/index.html

Biotech's Sparse Harvest
A Gap Between the Lab and the Dining Table

By Andrew Pollack
***The New York Times,* February 14, 2006**

At the dawn of the era of genetically engineered crops, scientists were envisioning all sorts of healthier and tastier foods, including cancer-fighting tomatoes, rot-resistant fruits, potatoes that would produce healthier French fries and even beans that would not cause flatulence.

But so far, most of the genetically modified crops have provided benefits mainly to farmers, by making it easier for them to control weeds and insects.

Now, millions of dollars later, the next generation of biotech crops—the first with direct benefits for consumers—is finally on the horizon. But the list does not include many of the products once envisioned.

Developing such crops has proved to be far from easy. Resistance to genetically modified foods, technical difficulties, legal and business obstacles and the ability to develop improved foods without genetic engineering have winnowed the pipeline.

"A lot of companies went into shell shock, I would say, in the past three, four years," said C. S. Prakash, director of plant biotechnology research at Tuskegee University. "Because of so much opposition, they've had to put a lot of projects on the shelf."

. . . Most of the genetically modified crops have provided benefits mainly to farmers

Developing nonallergenic products and other healthful crops has also proved to be difficult technically. "Changing the food composition is going to be far trickier than just introducing one gene to provide insect resistance," said Mr. Prakash, who has promoted agricultural biotechnology on behalf of the industry and the United States government.

In 2002, Eliot Herman and his colleagues got some attention when they engineered a soybean to make it less likely to cause an allergic reaction. But the soybean project was put aside because baby food companies, which he thought would want the soybeans for infant formula, instead are avoiding biotech crops, said Mr. Herman, a scientist with the Department of Agriculture.

In addition, he said, food companies feared lawsuits if some consumers developed allergic reactions to a product labeled as nonallergenic.

The next generation of these crops—particularly those that provide healthier or tastier food—could be important for gaining consumer acceptance of genetic engineering. The industry won a victory last week when a panel of the World Trade Organization ruled that the European Union had violated trade rules by halting approvals of new biotech crops. But the ruling is not expected to overcome the wariness of European consumers over biotech foods.

New crops are also important for the industry, which has been peddling the same two advantages—herbicide tolerance and insect resistance—for 10 years. "We haven't seen any fundamentally new traits in a while," said Michael Fernandez, executive director of the Pew Initiative on Food and Biotechnology, a nonprofit group.

Now, some new types of crops are appearing. Monsanto just won federal approval for a type of genetically engineered corn promoted as having greater nutritional value—albeit only

for pigs and poultry. The corn, possessing a bacterial gene, contains increased levels of lysine, an amino acid that is often provided to farm animals as a supplement.

Coming next, industry executives say, are soybean oils intended to yield healthier baked goods and fried foods. To keep soybean oil from turning rancid, the oil typically undergoes a process called hydrogenation. The process produces trans fatty acids, which are harmful and must be disclosed in food labels under new regulations.

Both Monsanto and DuPont, which owns the Pioneer Hi-Bred seed company, have developed soybeans with altered oil composition that, in some cases, do not require hydrogenation. Kellogg said in December that it would use the products, particularly Monsanto's, to remove trans fats from some of its products.

Monsanto's product, Vistive, and DuPont's, which is called Nutrium, were developed by conventional breeding. They are genetically engineered only in the sense that they have the gene that allows them to grow even when sprayed with the widely used herbicide Roundup.

But Monsanto and DuPont say the next generation of soybean, which would be able to eliminate trans fats in more foods, would probably require genetic engineering. Those products are expected in three to six years.

Beyond that, both companies said, would be soybeans high in omega-3 fatty acids, which are good for the heart and the brain. These are now derived largely from eating fish, which in turn get them by eating algae. Putting algae genes into soybeans could allow for soy oil that is rich in the fatty acids.

"Our hope is it is easier to formulate into food without it smelling or tasting fishy," said David M. Stark, vice president for consumer traits at Monsanto.

Other second-generation crops are also on the way. DuPont is trying to develop better tasting soy for use in products like protein bars.

Some efforts are under way to develop more nutritious crops for the world's least developed countries, led by what is termed golden rice, which contains the precursor of vitamin A. Vitamin A deficiency is a leading cause of blindness in certain poor countries.

There has been progress in crops able to withstand drought. While those would mainly benefit farmers, it would also help consumers in regions like Africa, where droughts bring famine.

Mr. Stark said Monsanto had not anticipated that use of genetic engineering would discourage food companies from using the company's soybeans. "I don't get many requests for 'Is this a G.M.O. or not?' " he said, using the abbreviation for genetically modified organism. "It's more 'Does the oil work?' "

Still, opposition by consumers and food companies has clearly forced big companies like Monsanto and DuPont to choose their projects carefully. It has also made it difficult for academic scientists and small start-ups, which typically provide much of the innovation in other fields, to obtain financing.

A Growing Focus on Consumer Benefits

HEALTHIER SOYBEAN OILS Soy oil is being modified to increase its stability and eliminate the need for hydrogenation — the process that produces unhealthy trans fats. Foods made with the modified oils would have reduced trans fats or none. Both Monsanto and DuPont recently introduced soybeans with healthier oils made by conventional breeding, which Kellogg plans to use for several products. In three to six years, similar products created by genetic engineering may be on the market.

IMPROVED SOY PROTEIN DuPont is working on improving the taste of soy protein used in protein bars and other products.

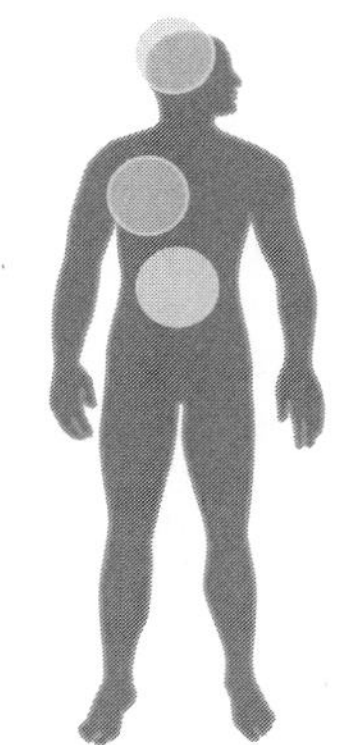

INCREASED OMEGA-3 FATTY ACIDS These chemicals, healthy for the heart and brain, come from fish, who get theirs from eating algae. Both Monsanto and DuPont are developing soybeans with the omega-3 fatty acid gene from algae.

MORE DIGESTIBLE AND NUTRITIOUS SORGHUM This product is being developed by DuPont with nonprofit groups. Sorghum is a staple food for many people in Africa, but it is hard to digest and not very nutritious.

NONALLERGENIC CROPS Genetic engineering might be used to make soy, wheat, peanuts and other crops that do not cause allergic reactions. Still in the early research phase.

ENHANCED RICE Vitamin A-enhanced rice, also known as golden rice, may prevent blindness caused by vitamin A deficiency, found in developing countries.

CANCER-FIGHTING TOMATO Research on tomatoes with higher levels of lycopene, an antioxidant, has slowed.

BETTER-TASTING PRODUCE Hardier tomatoes can ripen on the vine longer, producing superior taste.

Source: Companyreports

The New York Times

Avtar K. Handa, a professor at Purdue, said he had stopped work on a tomato he helped develop a few years ago that was rich in lycopene, a cancer-fighting substance. Genetically modified crops are not being brought to market and research funds have diminished, he said.

Still, opposition is not the only problem. Alan McHughen, a professor at the University of California, Riverside, said that for small companies and university researchers, the main obstacles were patent rights held by the big companies and the cost of taking a biotech crop through regulatory review. That has made it particularly difficult to apply genetic engineering to crops like fruits and vegetables, which have smaller sales than the major grain and oil crops.

Technical issues are another obstacle. While a single bacterial gene can provide herbicide resistance or insect resistance, changing the nutritional composition of crops sometimes requires several genes to alter the metabolism within a cell. That raises a greater risk of unintended effects, some experts say.

Enhanced crops must also meet the demands of farmers for high yields and of food companies for good taste and handling properties.

DuPont won approval for a soybean high in oleic acid, which could produce healthier oils, back in 1997. But instead of becoming a showcase of the consumer health benefits of genetic engineering, the crop is now used only to make industrial lubricants.

Erik Fyrwald, group vice president of DuPont's agriculture and nutrition division, said one reason the crop was not sold for use in food was that demand for healthier oils was not as great then as it is now. But other experts say there was another problem—foods made with the oil did not tastegood.

"The high-oleic oils are not very well received by the consumer," said Pamela White, a professor of food science and human nutrition at Iowa State University. Further, she predicted that soy oils containing the omega-3 fatty acids would be unstable, making them hard to use in fried foods.

William Freese, a research analyst at Friends of the Earth, which opposes genetically engineered crops, said genetic engineering had been oversold. "The facts show that conventional breeding is more successful at delivering crops with 'healthy traits' than genetic manipulation, despite all the hype from Monsanto and other biotech companies," he wrote in an e-mail message.

Scientists at the International Maize and Wheat Improvement Center in Mexico have already used conventional breeding to develop corn rich in lysine, similar to the new Monsanto product, he said.

The biotech companies concede that if improvements can be made conventionally, results would come quicker because such crops do not face regulatory scrutiny. Mr. Stark of Monsanto said that if his company could develop high-oleic soybeans using breeding, the product could reach the market in three years, rather than six for the genetically engineered version.

But in some cases, scientists and executives say, it is not possible to get a trait, like the omega-3 fatty acids, without using genes from another species. "With genetic engineering you can go further," said Mr. Fyrwald of DuPont.

Mr. Fernandez of the Pew Initiative said polls have shown that consumers seem to be receptive to genetically modified products that have direct benefits for them. But whether that would be enough to win wide acceptance of genetically engineered foods remains to be seen.

One issue is whether consumers would even know what they are eating. Right now, in the United States, genetically modified and conventional crops are typically mixed together, and food made from biotech crops is not labeled.

But it is likely that crops with consumer benefits would be segregated so farmers could charge more for them. And food companies are probably going to want to label them. But the labeling is likely to proclaim that the food has healthier oil or is better for the heart, rather than mention it was the product of genetic engineering.

In Europe, food containing genetically modified ingredients has to be labeled to that effect, but it is not clear whether the health aspects would be linked to genetic engineering on the label.

Chris Somerville, chief executive of Mendel Biotechnology, a small company developing drought-resistant crops, said acceptance would depend more on big food companies than consumers. Companies, he said, would not want to risk their brands by using biotech crops if they thought there was even a slight chance of consumer rejection.

"Really, they're the gatekeepers," said Mr. Somerville, who is also head of the plant biology department at the Carnegie Institution. "The consumers aren't going to have any choice before the brand companies think it's safe to go out."

Copyright 2006 The New York Times Company. Reprinted with permission.

In Review: Biotech's Sparse Harvest *A Gap Between the Lab and the Dining Table*

Testing Your Comprehension

1. In what way have most genetically modified crops provided benefits only to farmers?
2. What are the four major challenges in developing genetically modified crops?
3. In 2002, what were two of the food industry's reactions to efforts to engineer a soybean that would be less likely to cause an allergic reaction?
4. The genetic engineering industry won a victory as the result of a ruling by the World Trade Organization. What was the ruling?
5. Conventional breeding developed Monsanto's new product, Vistive, and Dupont's, Nutrium. In what way are they genetically engineered?
6. Name four new genetically engineered products that are currently under development.
7. Name four major challenges facing small companies and university researchers in developing new genetically engineered products.
8. Biotech companies concede that if breeding improvements can be made conventionally, results would come quicker. Why?

9. In the United States, food made from biotech crops is not labeled. According to the article, is it more likely that the labels will emphasize consumer benefits or mention that products were the result of genetic engineering?
10. According to Chris Somerville, who will drive acceptance of products containing genetically modified ingredients?

Weighing the Issues

1. Companies are using genetic engineering to incorporate different traits into consumer products. Examples include corn containing increased levels of lysine and soybean oils intended to yield healthier baked goods and fried foods. What types of traits, if any, would you like to see incorporated into consumer products?
2. What are the relative benefits and drawbacks of genetically engineered products?
3. Some efforts are under way to develop more nutritious and resilient crops for the world's least developed countries. Do you think genetic engineering is more justified for use in developing versus developed countries? Why or why not?
4. Why do you think consumers are not as interested in genetically engineered products as some companies originally expected them to be?
5. Do you think companies should be required to label genetically engineered products to explain that those products were the result of genetic engineering? Why or why not?

More to the Story

Organize a panel discussion on the pros and cons of genetically modified products for consumers. Identify three speakers presenting different perspectives: pro, con, and moderate. Provide a one-paragraph summary of a presentation and rebuttal for each participant.

Useful Websites

http://www.whybiotech.com

http://www.biotech-info.net

http://www.saveorganicfood.org/

Deals Turn Swaths of Timber Company Land Into Development-Free Areas

By Felicity Barringer
***The New York Times*, April 2, 2006**

BRITTONS NECK, S.C., March 28—Timber companies and conservation organizations have been working to arrange and announce a cascade of deals transferring large, unbroken swaths of forestland into the hands of government, nonprofit—or even commercial—groups that are committed to keeping them free from development.

On Tuesday, the International Paper Company announced it would receive $300 million in a deal arranged by the Nature Conservancy and the Conservation Fund for 217,000 acres in 10 states around the Southeast.

The largest single tract, an unkempt 25,668-acre peninsula between the Pee Dee and Little Pee Dee Rivers in South Carolina, will ideally revert to the cypress and longleaf-pine forest that once covered these sandy flatlands. The company also said it had sold 69,000 acres of forestland in Wisconsin for $83 million to the Nature Conservancy.

The third and largest deal is intended to preserve up to 400,000 acres of land near Moosehead Lake in central Maine. Financial and other details are still being worked out between the Plum Creek Timber Company, the Nature Conservancy and two regional conservation groups.

But for all the good news, celebrated by all sides, a stubborn fact remains: The nearly one million acres that have been preserved in these deals over the past two years, including a 257,000-acre tract in the Adirondacks, represent barely 2 percent of timber company lands that are coming on the market in the East.

And in many places like parts of North and South Carolina, conservation groups are competing for the land with developers who seem more determined than ever.

"Based on market components," said David Liebetreu, International Paper's vice president for forest resources, "our forestlands are worth a lot more to other people than they are to us."

Photographs by Stephen Morton for The New York Times

A bond issue in South Carolina helped ensure that the 13,281-acre Hamilton Ridge tract of land, owned by a series of timer companies, will be protected from development.

International Paper is not the only company making that calculation.

"The timber industry has taken nearly a century to assemble these large blocks of forestland," said Derb Carter, a lawyer with the Southern Environmental Law Center.

Craig Culp, who runs the Wilderness Society's eastern forest programs, said, "We're talking about forever altering the landscape."

The quick succession of sales, Mr. Carter said, provide a golden opportunity for conservation organizations—but the amount of state, federal and nonprofit money available is dwarfed by the amounts that can be offered by developers of residential communities, golf courses and hunting clubs. "The federal government is, for practical purposes, out of the conservation land acquisition business," Mr. Carter said.

An analysis of federal budget data by the Wilderness Society shows conservation financing—money available for conservation purchases either directly or through grants to states—has shrunk to about $140 million annually from more than $500 million in 2001.

At the moment, International Paper's land is the focus of attention. South Carolina, where two of its largest tracts going into conservation are located, expects to add 1 million residents to its population of 4.2 million in the next decade, said John E. Frampton, director of the state's Department of Natural Resources. There are 100 golf courses in the Myrtle Beach area, he added, and The Charleston Post recently reported that there were 134,000 building permits in Charleston County alone.

If the 39,000 acres in the two tracts—Woodbury in Brittons Neck and Hamilton Ridge on the other side of the state, near the Georgia border—were up for sale, Mr. Frampton

Nesting egrets and loblolly and longleaf pine cones at the Hamilton Ridge tract, near the Georgia border.

predicted, it would be bought up instantly and subdivided into hunting clubs and hobby farms and eventually second-home communities.

The state had been trying to acquire those two tracts from their succession of timber company owners for more than a decade. When International Paper made clear last summer that it would sell all or part of its 6.5 million acres of forestland around the East, the conservation groups brokered a deal; this week, the $32 million bond issue that cements it was signed by Gov. Mark Sanford, a Republican. And here, as with the other tracts, International Paper is guaranteed a supply of timber for its nearby mills for at least five years.

. . . The amount of state, federal and nonprofit money available is dwarfed by the amounts that can be offered by developers

The South Carolina bonds ensure that the state is the eventual heir to the Woodbury tract, where canebrakes and occasional skeletons of old cypress trees interrupt the not-so-neat rows of half-grown loblolly pines. Across the state, in one of South Carolina's fastest-growing regions, the 13,281-acre Hamilton Ridge tract, with its egret rookeries, will also be protected from development.

The tracts were chosen based on their ecological value, including the presence of endangered species, the stock of hardwoods and softwoods, and the proximity to other protected areas, said Mr. Liebetreu of International Paper; Steve McCormick, the Nature Conservancy's chairman; and Larry Selzer, who heads the Conservation Fund.

When the Nature Conservancy heard about the proposed sales by International Paper, "we laid out a long list of properties," totaling about a million acres, Mr. McCormick said. That wish list was pared down based on available financing and the company's preferences.

"It's a great example of private, public and nonprofit cooperation," said International Paper's chairman and chief executive, John Faraci.

The fact that the paper company retains the right to a supply of timber rankles a few environmentalists, but not Jamie Dozier, a biologist with the South Carolina Department of Natural Resources. Mr. Dozier grew up not far from the sandy second- and third-growth forest that thrives in the swampy areas between the Pee Dee River, which carries rich red soils from the Piedmont, and the blackish, acidic Little Pee Dee.

"They are the only folks who own very large pieces of land," Mr. Dozier said.

Referring to booming Myrtle Beach 40 miles away, Mr. Dozier said, "You'll have a 1,000-home subdivision pop up there in two weeks." While it will be a decade or more before that kind of development pressure arrives here in Marion County, he said, such deals are necessary to prepare for the future.

"You can't change concrete," Mr. Dozier said. "You can change a loblolly plantation."

Copyright 2006 The New York Times Company. Reprinted with permission.

In Review: Deals Turn Swaths of Timber Company Land Into Development-Free Areas

Testing Your Comprehension

1. What are the terms of the deal between the International Paper Company, the Nature Conservancy, and the Conservation Fund?
2. What is the goal of the largest deal, the 25,668-acre peninsula between the Pee Dee and Little Pee Dee Rivers?
3. The approximately 1 million acres that have been preserved in these deals over the past 2 years represent what percentage of timber company lands that are coming on the market in the East?
4. Provide two reasons why it is in International Paper's interest to sell some of their forestlands.
5. What is the major challenge preventing more of these types of deals from occurring?
6. How much conservation financing is currently available, and how does this compare to what was available in 2001?
7. What action cemented the deal that enabled South Carolina to acquire the Woodbury and Hamilton Ridge tracts?
8. According to what criteria were the protected tracts chosen?
9. When the Nature Conservancy heard about proposed sales by International Paper, they laid out a wish list of properties. What two factors played a role in paring down that list?
10. According to Jamie Dozier, why is it beneficial to strike a deal that enables the paper company to retain the right to a supply of timber?

Weighing the Issues

1. David Liebetreu of International Paper says, "Based on market components, our forestlands are worth a lot more to other people than they are to us." Do you think more opportunities to arrange these types of deals will emerge as the market continues to change, and if so, under what types of circumstances?
2. An analysis of federal budget data shows that conservation financing money has shrunk over the past five years. What more can be done to raise the funding necessary to complete these kinds of deals?
3. According to John Frampton of the South Carolina Department of Natural Resources, if the land in the Woodbury and Hamilton Ridge tracts were up for sale, it would be subdivided into hunting clubs and hobby farms and eventually second-home communities. What kind of impact would this have on the landscape and environment?
4. According to the article, the fact that the paper company retains the right to a supply of timber rankles a few environmentalists, but others (like Jamie Dozier) aren't upset by the arrangement. What is your opinion?
5. Beyond the type of deal described in this article, what other types of opportunities exist to protect land from development?

More to the Story

Develop a management plan for the preserved tracts that attempts to connect them to other protected ecosystems in the area, preserving ecological corridors as well as facilitating low-impact recreation (e.g., hiking, bird watching) and other appropriate uses of the land given its preserved status. What areas should be placed entirely off-limits? How might limited ecotourism activities affect the local or state economy? What areas of the economy would be affected? Where might logging occur?

Useful Websites

http://www.thestate.com/mld/thestate/14210276.htm?template=contentModules/printstory.jsp

http://www.dnr.sc.gov

http://www.nature.org/wherewework/northamerica/states/southcarolina

Americans Are Cautiously Open to Gas Tax Rise, Poll Shows

By Louis Uchitelle and Megan Thee; Marina Stefan contributed reporting for this article.
***The New York Times,* February 28, 2006**

The New York Times | CBS NEWS Poll

Americans on Gasoline Taxes

Would you favor or oppose an increased federal tax on gasoline?

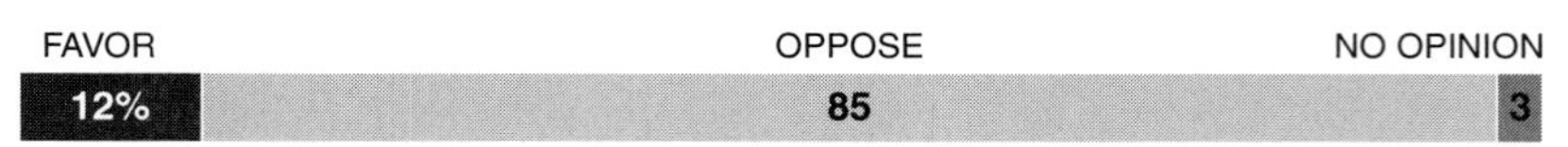

What if the increased tax on gasoline would reduce the United States' dependence on foreign oil, then would you favor or oppose an increased federal tax on gasoline?

What if the increased tax on gasoline would cut down on energy consumption and reduce global warming, then would you favor or oppose an increased federal tax on gasoline?

FAVOR	OPPOSE	NO OPINION
59%	34	7

Based on nationwide telephone interviews conducted by The New York Times and CBS News Feb. 22-26 with 1,018 adults.

The New York Times

Americans are overwhelmingly opposed to a higher federal gasoline tax, but a significant number would go along with an increase if it reduced global warming or made the United States less dependent on foreign oil, according to the latest New York Times/CBS News poll.

The nationwide telephone poll, conducted Wednesday through Sunday, suggested that a gasoline tax increase that brought measurable results would be acceptable to a majority of Americans.

Neither the Bush administration nor Democratic Party leaders make that distinction. Both are opposed to increasing the gasoline tax as a means of discouraging consumption, although President Bush, in recent speeches, has called for the development of alternative energy to reduce dependence on foreign oil.

Eighty-five percent of the 1,018 adults polled opposed an increase in the federal gasoline tax, suggesting that politicians have good reason to steer away from so unpopular a measure. But 55 percent said they would support an increase in the tax, which has been 18.4 cents a gallon since 1993, if it did in fact reduce dependence on foreign oil. Fifty-nine percent were in favor if the result was less gasoline consumption and less global warming. The margin of sampling error is plus or minus three percentage points.

"The tax would have to be earmarked for certain specific projects," one of the people polled, Rich Arnold, 54, a Republican who teaches criminal justice at Louisiana State University, said in a follow-up interview. He added, "If it was a tax that would sponsor research for fuel cells or alternative fuel sources, I could buy that."

Some people were concerned that a higher gasoline tax would find its way into what they considered the wrong hands—unless it were directed to a specific use, just as the current 18.4-cent tax is channeled into highway maintenance and construction. Lisa Fisher, a 36-year-old yoga instructor in Chicago who described herself as a Democrat, wants any additional revenue earmarked.

"If the tax is increased and oil companies reap the benefit, I would be against it," Ms. Fisher said. "But if the tax money went to the development of electric cars, I would favor the higher tax. It is important we are not dependent on foreign oil. We are over there fighting because we are dependent."

Twenty-four percent of those polled said they would support a higher federal gasoline tax if the new revenue was used to help fight terrorism, and 28 percent would go along with a gasoline tax increase if, as an offset, their income taxes or payroll taxes were lowered. Betty Forde, 46, an administrator for the New York City Police Department, fell into this category.

"Taxes in general are very high," said Ms. Forde, who earns $50,000 a year and does not own a car or drive.

"This may be a bit selfish on my part," she said, "but I use taxis a lot, and if the gasoline tax goes up, they are going to charge me more. There must be some offset."

Many mainstream economists believe that a shift that raises the gasoline tax while lowering income-based taxes is the most efficient way to reduce consumption. It might require a $1-a-gallon increase in the tax phased in over five years, said Severin Borenstein, director of an energy institute at the University of California, Berkeley.

Because increasing the gas tax is regressive, falling hardest on those who can least afford it, Mr. Borenstein would offset the bite by lowering income taxes in a way that would "make most middle and lower income people better off." But they would end up driving less because of the rising cost of gasoline, some economists believe. By Mr. Borenstein's calculation, a 10 percent increase in the price of gasoline reduces consumption by 6 to 8 percent "over the long run."

. . . a gasoline tax increase that brought measurable results would be acceptable to a majority of Americans

A higher tax might have carried some weight with Rosalee Evans of Norfolk, Va., before the price of gasoline broke above $2 a gallon last March and peaked in early September, a week after Hurricane Katrina, at $3.07. The nationwide average for regular-grade gasoline has since fallen, to $2.24 a gallon last week, the Energy Department reports. That includes the federal tax and 21.04 cents, on average, in various state and local taxes.

"At least get the price down to $2 or $2.10 a gallon," said Ms. Evans, who works for a firm that brings entertainers to the Tidewater area and who drives a lot, recruiting talent. "I could live with that price, and then I would consider raising the federal gasoline tax to combat global warming, but a reasonable tax increase, not a big one."

A falling gasoline price, however, would undermine the pressure on people to buy fuel-efficient cars and move closer to their work, reducing their commute.

The problem is that the transition would be painful, even punitive, said Ashok Gupta, an economist at the Natural Resources Defense Council, an environmental advocacy group. To speed things, he would funnel some of the additional tax revenue to manufacturers "as an incentive to offer more efficient vehicles, like hybrid cars."

Vincent Bussey, who earns $34,000 maintaining computers at a bank in Los Angeles, agreed. Like most of those polled, he opposed a higher gasoline tax unless it were used to develop "new technologies that made the nation somewhat more self-reliant."

Copyright 2006 The New York Times Company. Reprinted with permission.

In Review: Americans Are Cautiously Open to Gas Tax Rise, Poll Shows

Testing Your Comprehension

1. What was the overarching finding of the New York Times/CBS News poll?
2. Under what specific circumstances would a gasoline tax increase be acceptable to a majority of Americans?
3. The Bush administration and Democratic Party leaders are opposed to increasing the gasoline tax as a means of discouraging consumption. What has President Bush called for in recent speeches?
4. What percentage of adults polled a) opposed an increase in the federal gasoline tax, b) said they would support an increase if it did reduce dependence on foreign oil, and c) were in favor if the result was less gasoline consumption and less global warming?
5. Some respondents said they would support an increased tax if it were earmarked for certain purposes. Name two purposes that were cited.
6. According to one respondent, what might happen if the increased tax were not earmarked?
7. Increasing the gas tax is regressive, falling hardest on those who can least afford it. How could this impact be avoided?
8. According to Severin Borenstein, to what extent would a 10 percent increase in the price of gasoline reduce consumption?
9. One respondent would need to see the price of gas decrease before supporting an increase in the gasoline tax. What impact would a falling gas price have on consumption?
10. What does Ashok Gupta of the Natural Resources Defense Council recommend as a way to speed up the transition to reduced gasoline consumption?

Weighing the Issues

1. According to the New York Times/CBS News poll, a gasoline tax that brought measurable results would appeal to many Americans. What do you think is meant by "measurable results"?

2. Some respondents said they would support an increased tax if it were earmarked for certain specific projects. What earmarked projects would make the greatest progress in reducing the country's demand on foreign oil? What projects would help reduce global warming?
3. If the gas tax were increased to an intolerably high level, what steps would you take to reduce your own consumption of gas? What other externally imposed constraints would lead you to make such changes?
4. Some respondents said they would favor an increased tax only if it were offset by another action, e.g., lowering other taxes. What other offsets might be acceptable?
5. In what ways might higher gas prices lead to a decrease in oil consumption?

More to the Story

If you were interviewed as part of the New York Times/CBS News poll, how would you respond to the questions about your potential support of an increased gasoline tax? Why? What do you think should be the approximate price of gas necessary to encourage conservation, and why? Do you agree with the proposal to offset a gas tax increase with increased payroll (or other taxes)? Why or why not? Lay out a three-pronged program using taxes, incentives, or other measures to encourage gas conservation, and explain your rationale.

Interpreting Graphs and Data

1. Based on the data shown, which potential outcome—reduced dependence on foreign oil or reduced global warming—would respondents favor more greatly as the result of increased gas taxes?
2. By how much of a margin would respondents prefer to see both outcomes combined, as opposed to the most favorable outcome?

Useful Websites

http://sfgate.com/cgi-bin/article.cgi?f=/c/a/2006/05/10/MNGQHIOUJ71.DTL

http://economistsview.typepad.com/economistsview/2006/06/a_gas_tax_with_.html

http://www.env-econ.net/2005/07/fuel_efficiency.html

The New Face Of an Oil Giant
Exxon Mobil Style Shifts a Bit

By Jad Mouawad
***The New York Times,* March 30, 2006**

If Rex W. Tillerson has his way, Exxon Mobil will no longer be the oil company that environmentalists love to hate.

Since taking over as Exxon's chairman three months ago from Lee R. Raymond, his abrasive predecessor who dismissed fears of global warming and branded environmental activists "extremists," Mr. Tillerson has gone out of his way to soften Exxon's public stance on climate change.

"We recognize that climate change is a serious issue," Mr. Tillerson said during a 50-minute interview last week, pointing to a recent company report that acknowledged the link between the consumption of fossil fuels and rising global temperatures. "We recognize that greenhouse gas emissions are one of the factors affecting climate change."

But despite the shift in style to a less adversarial tone, the substance of Exxon's position has not changed with the new chairman. The company said the recent report only clarified its long-held position on global warming. Indeed, Mr. Tillerson noted that he, like Mr. Raymond before him, remained convinced that there was "still significant uncertainty around all of the factors that affect climate change."

To Fadel Gheit, a longtime industry analyst at Oppenheimer & Company in New York, Mr. Tillerson certainly presents a kinder, gentler face for Exxon. But in the end, Mr. Gheit cautioned, do not expect much difference between Mr. Tillerson and Mr. Raymond.

Photograph by Jerry W. Hoefer for The New York Times

"What we do brings good things to people," Rex W. Tillerson, Exxon Mobil's chairman, said of the company's role in society.

"It's the same old wine in a new bottle," he said. "You can't expect a company this size to change on a dime, but you might see changes in how it projects its image to the public, to its clients."

Mr. Tillerson sees Exxon's future as still firmly tied to oil and natural gas

"Lee was impatient," he added. "Rex is firm, but with a smile."

Mr. Tillerson, who succeeded Mr. Raymond in January, said he saw no reason for any sharp departure in strategy. Exxon's business is about increasing oil and gas supplies to consumers, he said, not chasing alternatives that offer little prospect of replacing the fossil fuels that he views as the only realistic way to meet the world's huge and growing demand for energy.

In contrast to rivals at BP and Royal Dutch Shell, which plan to invest billions of dollars in the next decade to develop renewable energy sources like wind and solar power, Mr. Tillerson sees Exxon's future as still firmly tied to oil and natural gas.

The answer to today's high prices? "More supplies." President Bush's reference to America's "addiction to oil?" "An unfortunate choice of words." Exxon's role in society? "A good business, and what we do brings good things to people."

"To say suddenly that there is something wrong about that," he said, "I can't connect with that."

With 30 years at Exxon, Mr. Tillerson has taken over the company at a time when the oil industry faces formidable new challenges. Not since the 1980's has there been as much talk about energy costs and the nation's dependence on oil.

After nearly two years of high energy prices, oil companies are facing public discontent at $2.50-a-gallon gasoline and political pressure over the companies' record profits. Mr. Tillerson said the situation offered him an opportunity to better explain his company's position.

Exxon Mobil Leads the Pack

Exxon Mobil's discipline has paid off and its chairman, Rex Tillerson, sees no reason to change a strategy that has proved highly successful.

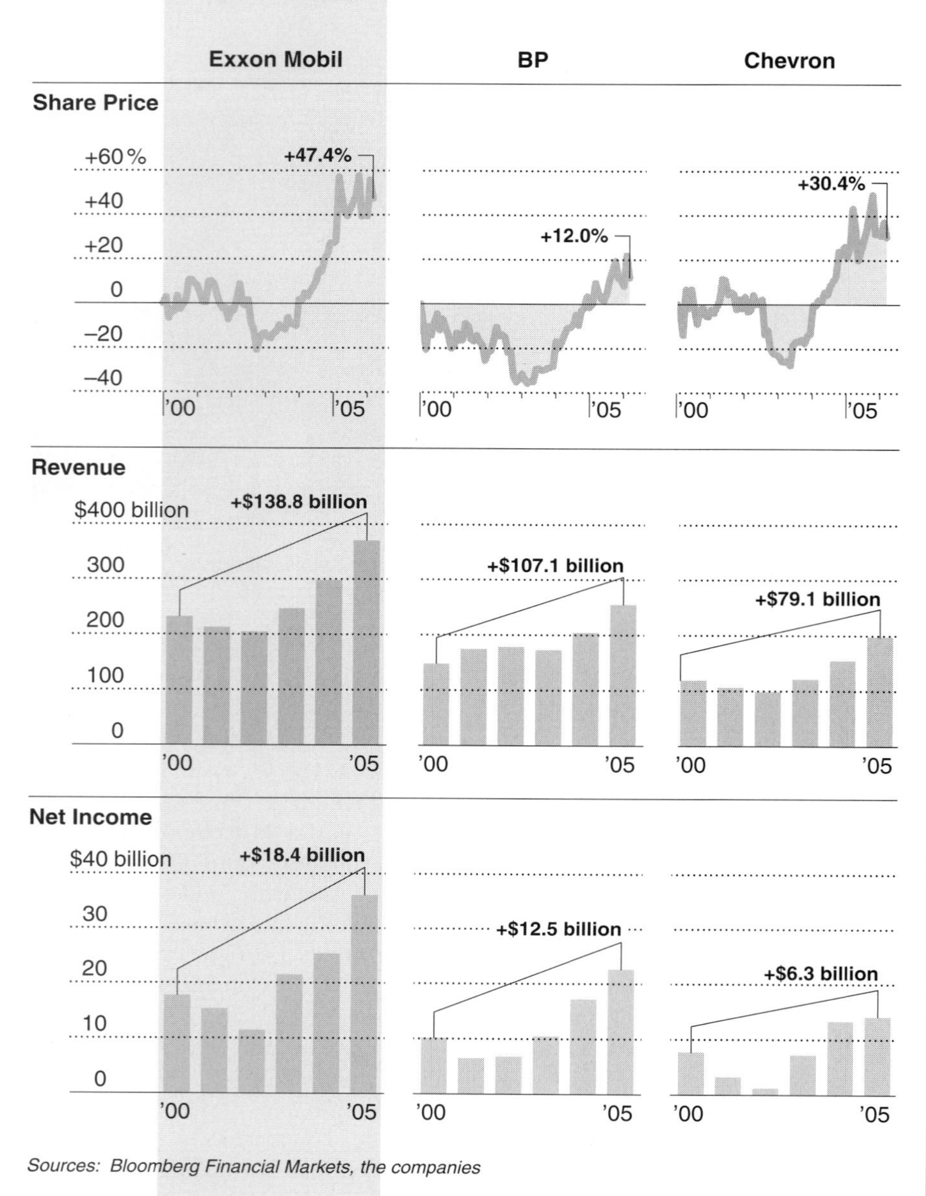

Sources: Bloomberg Financial Markets, the companies

The New York Times

"The only thing I've said to people will change, maybe, is the management style, the way I communicate," Mr. Tillerson said. "We're all individuals. Lee Raymond is Lee Raymond. He has his style. I am Rex Tillerson and I have my style."

Whatever Mr. Raymond's legacy on the environmental front, there's no arguing with Exxon's financial success. He pulled the company far ahead of rivals by engineering the 1999 merger with Mobil that partly recreated the original Standard Oil trust.

Exxon is now the world's largest publicly traded oil and gas producer. Last year, its net income surged to $36.1 billion, the highest for any American corporation and a 43 percent jump from the previous year. That is a legacy Mr. Tillerson is proud to defend.

But to its many critics, Exxon, based in Irving, Tex., is locked in an increasingly frustrating race for additional oil supplies and is failing to help develop alternative fuels, curb consumption and act on the real threat of global warming.

"They have to be part of the solution," said Kert Davies, a research director at Greenpeace. "They have too much money; they are too powerful. Without Exxon pulling with the rest of the world, it will take longer to solve global warming."

For Shawnee Hoover, the campaign director of Exxpose Exxon, a coalition of the nation's leading environmental groups, including Greenpeace and the Sierra Club, "Exxon has this prehistoric culture."

She added: "They dig their heels in."

But at Exxon, executives see very little reason to alter a course that has proved exceptionally profitable.

In a capital-intensive business, the company's obsession about costs has allowed it to outperform all its rivals. Its rate of return on capital employed, which the company says is the best indication of performance and cost management, reached 31 percent last year. The second-highest return among the giant oil companies, BP's, was 20 percent.

"Exxon has really been about discipline," said Daniel L. Barcelo, an analyst at Banc of America Securities. "What Exxon brings to the table is their balance sheet, the technical expertise, and their operational management and development. That's where they shine."

But he said the company's conservative management also had a flip side. "Others have been more willing to take risks," Mr. Barcelo said. "Some say Exxon is actually being blind and missing out on huge opportunities for growth."

Indeed, oil analysts argue that the company has been plowing too little money back into finding hydrocarbons while giving too much back to shareholders. Oil and gas production as well as reserves have remained mostly flat for the last five years. Last year, Exxon paid $23.2 billion in dividends and share buybacks, more than the $17.7 billion it spent on exploration and development.

A native of Wichita Falls, Tex., Mr. Tillerson, who turned 54 this month, joined Exxon in 1975 as a production engineer after graduating from the University of Texas with a degree in civil engineering. He later ran some of Exxon's American operations. In the early 1990's, he was responsible for negotiating the company's investments in Sakhalin Island in Russia, as well as in the Caspian Sea.

Since he started at Exxon, the energy business has changed radically. Easy-to-find oil has been mostly found, opportunities for new resources are scarcer, competition is rising and governments are tightening the screws on international oil companies.

But after taking over as chairman, Mr. Tillerson has already scored two major coups: gaining access to the world's fourth-largest oil field, in the United Arab Emirates, and prevailing in a five-year-old dispute over the development of Indonesia's largest untapped oil reserves.

Mr. Tillerson met with each country's leaders to break deadlocked talks or make a final pitch for his company. In Indonesia, the government fired the head of the national oil company, who opposed Exxon. His successor quickly signed a deal.

But if both agreements proved a success for Mr. Tillerson, they also mask a starker reality for oil companies: their access to the world's top hydrocarbon deposits is more limited than ever. At Exxon, the problem is magnified by the company's size. Each year, its geologists must find huge amounts of oil and gas—nearly 1.5 billion barrels—just to replace the company's production of about 4 million barrels a day.

The model for Exxon's expansion was perhaps best displayed in Qatar, a small Persian Gulf state holding the world's third-largest natural gas deposits, after Russia and Iran. In the early 1990's, Exxon approached the Qatari government with an offer to serve as a joint partner. Today, Exxon is the largest foreign investor in Qatar and the nation is on track to become the world's leading liquefied natural gas producer.

"We are looking for the large opportunities," Mr. Tillerson said.

Referring to Qatar, Mr. Tillerson said "that approach can be replicated around the world."

But can it? Recent setbacks in Venezuela and Russia suggest the obstacles are multiplying. After briefly welcoming foreign oil producers, Russia has now mostly shut the door to new foreign investment. In Venezuela, Exxon is battling the demands of President Hugo Chávez's nationalist government, which wants to increase royalties and other taxes on foreign investors. But rather than give in and set a precedent, the company prefers to scale back its investments or shut fields.

Still, Mr. Tillerson insisted that Exxon was not constrained by a lack of prospects or partners. "There are other opportunities," he said, "in the Middle East, in the Caspian, in other parts of the world where we will continue to take the same approach."

This month, at Exxon's annual session for Wall Street analysts, top executives outlined 22 major projects over the next three years, from Angola to Norway, Malaysia to the North Sea. For 2009 and beyond, they identified another 32 prospects.

To develop these projects, the company plans to increase its capital spending to $20 billion a year by the end of the decade. Exxon hopes to increase its oil and natural gas production to five million barrels a day by 2010 and lift its daily capacity by a total of two million barrels after 2015.

Dismissing the view that the world is running out of oil, Mr. Tillerson said there was still plenty more to be found to meet what Exxon expects will be a 50 percent rise in global energy demand by 2030.

At the same time, he defended Exxon's record of investing in research for alternative fuels, citing a 10-year, $100 million contribution to the Global Climate and Energy Project at Stanford, which focuses on long-term technological research. "We are going to continue to use fossil fuels," he said. "We are looking for the fundamental changes, but that's decades away. The question is, What are we going to do in the meantime?"

Three months into the job, the changes at the helm of Exxon are mostly evident in small, impressionistic touches.

At a refiners' conference in Salt Lake City last week, for example, Mr. Tillerson urged other managers to get the industry's message out by, among other things, attending Rotary Club and PTA meetings.

And at a recent news conference, he displayed a lighter touch that one rarely associates with Exxon executives. In reply to a question about what he thought would happen to oil prices this year, for example, Mr. Tillerson offered this response: "If I knew, I'd be living on a Caribbean island with my flip-flops and a laptop, working just two hours a day."

Copyright 2006 The New York Times Company. Reprinted with permission.

In Review: The New Face Of an Oil Giant
Exxon Mobil Style Shifts a Bit

Testing Your Comprehension

1. What has Rex Tillerson done since taking over as Exxon's chairman that represents a shift in style to a less adversarial tone?
2. What is Rex Tillerson's stated position on global warming?
3. According to Mr. Tillerson, what is Exxon's strategy regarding gas supply and alternative fuels?
4. What steps are BP and Royal Dutch Shell taking to find alternatives to fossil fuels?
5. What was Exxon's rate of return on capital employed last year, and how does it compare to BP, which had the second-highest return among the giant oil companies?
6. How has the energy business changed since Mr. Tillerson joined Exxon in 1975?
7. After taking over as chairman, Mr. Tillerson has had two major successes. What are they?
8. Describe Exxon's relationship with Qatar as a model for expansion.
9. Exxon is planning 22 major projects by the end of the decade. What steps will they have to take to develop these projects?
10. What has Exxon done to invest in research for alternative fuels?

Weighing the Issues

1. According to Mr. Tillerson, there is still a significant amount of uncertainty about what factors affect climate change. Do you agree? Why or why not?
2. While Mr. Tillerson's position on climate change may be similar to his predecessor's, his style is different. How might this difference in style affect the ways in which Exxon is perceived in terms of its environmental performance?
3. Why do you think BP and Royal Dutch Shell are moving more aggressively toward alternative energy sources while Exxon's future is still firmly tied to oil and natural gas?
4. According to Daniel Barcelo of Bank of America Securities, "Others have been more willing to take risks. Some say Exxon is actually being blind and missing out on huge opportunities for growth." If you were a special advisor to Mr. Tillerson, how would you advise him on the extent to which the company should take new risks, and in what ways to do so?
5. Exxon expects that there will be a 50 percent rise in global energy demand by 2030. Do you think it will be possible to meet that demand, and if so, how?

More to the Story

BP and Royal Dutch Shell are pursuing an approach to oil conservation and public relations that is entirely different from that of Exxon Mobil. Lay out the key elements of each company's strategy, with attention to the practices they are undertaking, the resources they are developing, the probable outcome, and the economics of their approach. Evaluate the three programs in comparison to one another with regard to these criteria.

Interpreting Graphs and Data

1. From 2000–2005, what was Exxon Mobil's increase in share price as opposed to BP and Chevron's?
2. As of 2005, what was Exxon Mobil's revenue as compared to BP's and Chevron's?
3. From 2000–2005, what was Exxon Mobil's increase in revenue as opposed to BP and Chevron's?
4. As of 2005, what was Exxon Mobil's net income as compared to BP's and Chevron's?
5. From 2000–2005, what was Exxon Mobil's increase in net income as opposed to BP and Chevron's?
6. Which of the three companies has shown the greatest stability in net income from 2000–2005?

Useful Websites

http://www.iht.com/articles/2006/06/08/business/gas.php

http://www.exxonmobil.com/corporate/

http://www.bp.com/home.do?categoryId=1

http://www.shell.com

Corn Farmers Smile as Ethanol Prices Rise, but Experts on Food Supplies Worry

By Matthew L. Wald
***The New York Times,* January 16, 2006**

SIOUX CENTER, Iowa, Jan. 11—Early every winter here, farmers make their best guesses about how much food the world will demand in the coming year, and then decide how many acres of corn to plant, and how many of soybeans.

But this year is different. Now it is not just the demand for food that is driving the decision, it is also the demand for ethanol, the fuel that is made from corn.

Some states are requiring that ethanol be blended in small amounts with gasoline to comply with anti-pollution laws. High oil prices are dragging corn prices up with them, as the value of ethanol is pushed up by the value of the fuel it replaces.

"We're leaning more toward corn," said Garold Den Herder, a farmer who cultivates 2,400 acres in a combination of corn and soybeans and is on the board of directors of the Siouxland Energy and Livestock Cooperative, which opened an ethanol plant here in late 2001. Last year a bushel was selling for about $2 here, but near the plant it was about 10 cents higher.

Farmers expect it to go higher soon if oil prices stay high. Ethanol was up to $1.75 a gallon, last year, from just over $1 the year before.

The rising corn prices may be good news for farmers, but they are worrying some food planners.

"We're putting the supermarket in competition with the corner filling station for the output of the farm," said Lester R. Brown, an agriculture expert in Washington, D.C., and president of the Earth Policy Institute. Farms cannot feed all the world's people and its motor vehicles as well, Mr. Brown said, and the result is that more people will go hungry.

Others say that the price of goods that have corn as an ingredient, including foods like potato chips or Danish pastries, will rise.

But Robert C. Brown, a professor of mechanical engineering at Iowa State University and a specialist in agricultural engineering, said the use of corn for nonfood purposes sounded harsher than it was. "The impression is that we're taking food out of the mouths of babes," Professor Brown said. In fact, corn grown in Iowa is used mostly to feed farm animals or make corn syrup for processed foods.

And Bernie Punt, the general manager of the Siouxland plant, said, "It's not as big a loss as what it seems like," pointing out that the corn remnants that come out of the other end of the plant were used for animal feed.

A global shift to farm-based fuel could reduce the need for oil and slow climate change. But Lester Brown is not alone in worrying about the effect on world hunger. For 20 years, the International Food Policy Research Institute, a nonprofit group in Washington, has maintained a computer model to predict food supplies, based on population changes, farm policies and other factors.

Until now, the institute's analysis had included the price of oil and natural gas only as a factor in production costs, including the price of making fertilizer, running a tractor or hauling food to markets. But last year, after Joachim von Braun, the director of the institute, went to Brazil and India, both of which make vehicle fuel from plants, he told his economists to change the model, taking into account the demand for energy from farm products.

Even a small shift could have big effects, Mr. von Braun said, because "the mouth of your car is a monster compared to your family's stomach needs."

". . . The mouth of your car is a monster compared to your family's stomach needs"

"I do not just expect somewhat higher food prices, but new instability as well," he said in an interview. "In the future, instability of energy prices will be translated into instability in food prices."

Gustavo Best, the energy coordinator at the United Nations Food and Agriculture Organization, said growing crops for energy could provide new opportunities for small farmers around the world and finance the development of roads and other valuable infrastructure in poor rural areas.

But, Mr. Best added, "definitely there is a danger that the competition can hit food security and food availability."

Some experts scoff at the idea of corn shortages, but others say it is possible. Wendy K. Wintersteen, the dean of the College of Agriculture at Iowa State University, said that possibly as early as this summer, "we will have areas of the state we would call corn deficient," because there will not be enough for livestock feed—the biggest use of corn here—and ethanol plants.

"It's a hard thing to imagine in Iowa," Ms. Wintersteen said. Eventually, experts say, American corn exports could fall.

Nationwide, the use of corn for energy could result in farmers' planting more of it and less wheat and cotton, said Keith J. Collins, chief economist of the Department of Agriculture. But the United States is paying farmers not to grow crops on 35 million acres, to prop up the value of corn, he said, and

Photograph by Dave Eggen for The New York Times

In front of corn piled 35 feet high, Bernie Punt, left, manager of the ethanol plant in Sioux Center, Iowa, talked with Kent Pruismann, a board member of the group running the plant.

much of that land could come back into production.

A change is under way that experts say will tightly tie the price of crops to the price of oil: ethanol plants are multiplying.

Iowa has 19 ethanol plants now and will have 27 by the end of the year, said Mr. Punt, a former president of the Iowa Renewable Fuels Association. The Siouxland Energy and Livestock Cooperative showed a $6 million profit for 2005, Mr. Den Herder said, driven in part by the price of ethanol.

Many farmers here in the corn belt say they have the ability to grow the material for vast amounts of fuel. Another biofuel is a diesel substitute made from soybeans, which still leaves about 80 percent of the bean for cattle feed, advocates say.

Joe Jobe, executive director of the National Biodiesel Board, a trade group, predicted that more demand for soy oil as a diesel substitute would force production of meal, pushing down its price and thus making cattle feed cheaper.

"I think there's a historical shift under way, not to grow more crops for energy and less for food, but to grow more for both," Mr. Jobe said.

Nick Young, the president of an agriculture consulting firm, Promar, in Alexandria, Va., pointed out that corn products have been used for nonfood purposes for years, including to make fluids used to help drill oil wells. Mr. Young said it was an exaggeration to say that nonfood use of crops will make the world's poor go hungry, but he added that the use of vegetable oil as a substitute for diesel fuel had already driven up the price of canola oil.

"These markets are linked," Mr. Young said. "Inevitably, there's going to be some interaction on food prices."

Copyright 2006 The New York Times Company. Reprinted with permission.

In Review: Corn Farmers Smile as Ethanol Prices Rise, but Experts on Food Supplies Worry

Testing Your Comprehension

1. In addition to the demand for food, what other factor is now driving farmers' decisions on how many acres of corn to plant?
2. What effect are high oil prices having on corn prices, and why?
3. According to Lester Brown, how will increasing use of corn for ethanol affect world hunger?

4. What are the two most common uses of corn in Iowa?
5. Name two major impacts of a global shift to farm-based fuel.
6. According to Joachim von Braun, how will a shift to ethanol affect energy and food prices?
7. How might growing crops for energy benefit small farmers and poor rural areas around the world?
8. In the U.S., how might the use of corn for energy affect the volume and type of crops that farmers plant?
9. What is the U.S. doing to "prop up the price of corn," and how might that change with the increased use of corn for energy?
10. Another biofuel is a diesel substitute made from soybeans. According to Joe Jobe, how would increased demand for soy oil affect the production and price of meal and cattle feed?

Weighing the Issues

1. Lester Brown believes that farms cannot feed all the world's people and its motor vehicles as well, and the result is that more people will go hungry. Do you agree? Why or why not?
2. The United States is currently paying farmers not to grow crops on 35 million acres. Given what you have learned from the article about the need for food and the economics of corn and ethanol, does this surprise you? Why or why not?
3. A global shift to farm-based fuel could reduce the need for oil and slow climate change. What other actions, if any, could accomplish the same goals?
4. Ultimately, greater use of ethanol could affect food prices. Do you think it's worth it? Why or why not?
5. If you were a farmer in Iowa, would you change your farming practices given the increased demand for biofuels, and if so, in what ways?

More to the Story

Where do you stand on the extent to which increased ethanol production will threaten the world's food supply in terms of:

- The actual depletion of global food sources that could occur?
- Alternative fuel sources other than ethanol?
- The relative impact on developed vs. developing countries (and the ethics thereof)?
- The future livelihoods of U.S. farmers?

Useful Websites

http://www.energybulletin.net/17036.html

http://www.earth-policy.org/Indicators/Grain/index.htm

http://resourceinsights.blogspot.com/2006/05/newest-guest-at-your-dinner-table-your.html

Gualeguaychú Journal
A Back-Fence Dispute Crosses an International Border

By Larry Rohter
***The New York Times,* February 13, 2006**

GUALEGUAYCHÚ, Argentina—For Argentines, few traditions are more treasured at this time of year than a relaxed beach vacation, preferably in neighboring Uruguay. But the residents of this border town are risking their countrymen's wrath by blocking highways to Uruguay to protest the construction of a pair of paper mills there that they say will pollute the river that forms the frontier between the countries.

Just east of here, several dozen demonstrators, some playing cards, others sipping bitter maté tea from gourds or roasting sausages on grills, sat in the shade of a red cargo truck and a tractor that serve as a roadblock. "No to the paper mills, yes to life," proclaimed their bumper stickers and the banners they had hung from the truck.

"The Uruguayans have no right to poison a river that belongs to all of us on both sides," said José Pouler, the owner of a pizzeria here. "These projects are going to damage agriculture and kill off tourism, all for the benefit of a couple of foreign companies that don't care about the people of this region."

The paper mills—one owned by a Finnish-Swedish consortium, the other by a Spanish company—are being built on the riverbank in the Uruguayan town of Fray Bentos. They represent an investment of more than $1.9 billion, the largest in Uruguay's history, and are expected to produce more than 1.5 million tons of cellulose for export each year.

The road blockages here began just before the new year, after the residents of this town of 80,000 expressed frustration that their complaints were being ignored in both capitals. They accuse Uruguay of violating a treaty that governs use of the river, and are irritated that their own president, Néstor Kirchner, has not been more energetic in opposing the projects.

Initially, the protesters announced in advance where and when they would block highways and for how long, allowing vacationers to adjust their schedules. But the picketers have raised the ante, now acting without warning and not telling motorists how long the blockade will last.

The environmental group Greenpeace has also led protests aboard boats in the middle of the river. But spokesmen for the paper companies say that the factories will meet the demanding environmental standards of the European Union and will employ technology that reduces pollution to a minimum.

Though the two presidents have recently talked by phone about the standoff, they seem reluctant to make concessions that may offend their supporters

Some of the vacationers who have come long distances from the interior of Argentina, only to be turned back here or at two other border crossings north of here, have cursed the protesters and refused to take the pamphlets they are handing out. But the opponents of the paper mills show little sympathy for them.

"Our health and well-being are more important than their being able to spend their summer vacation on a beach in Uruguay," said Daniel Frutos, a physician here.

Luis Molivuevo, one of the boycott organizers, added, "We've asked other Argentines not to spend their summer in Uruguay, but if they don't want to help, then we have to make our boycott obligatory."

Commerce among the four countries that make up South America's Mercosur customs union is also suffering, and that has led Uruguayan authorities to charge that the promise of free movement in the group's founding charter is being violated. Trucks from Chile carrying mill equipment were forced to turn back, and on both sides of the border, drivers of other vehicles laden with cargoes of perishable food and machinery have been camped out, sometimes for days and with little money for food, waiting for the roadblock to be lifted.

Across the river, in sleepy Fray Bentos, sentiment is just as strong in favor of the projects. The town has been "economically dead" since a meat processing plant closed more than 20 years ago, said Dani Bazán, a commercial photographer there who welcomes the 2,000 new jobs and the revival of business activity the mills will bring.

"It's not that we like the idea of the mills so much as that we welcome the jobs, and well-paying ones, at that," said Sandra Caballero, a 35-year-old cook who is taking a course to become

Photograph by Pablo Cabado for The New York Times

Holding an Argentine flag, a protester in Gualeguaychú blocked a highway leading to Uruguay.

Gualeguaychú residents say two paper mills will harm their river.

a solderer in hopes of getting a job at the plant owned by the Finnish-Swedish consortium. "There will undoubtedly be some pollution, but we have faith that our government will be able to control emissions and punish the companies if they do something wrong."

For Uruguayans, the dispute has also become a matter of sovereignty and national pride. Their country was created 180 years ago as a buffer between Brazil and Argentina, and throughout their history they have often complained of being bullied and scorned by their much larger neighbors across the River Plate estuary, with whom they share a similar accent and culture.

Uruguay has recently expressed dissatisfaction with its secondary role in the Mercosur trade group and with the conduct of its neighbors. The left-leaning government of Tabaré Vázquez, which took office in March 2005, as a result has recently expressed interest in negotiating a free trade agreement with the United States; if reached, it would surely be a death blow to Mercosur.

Mr. Kirchner initially declared that stopping construction of the paper mills was "a national cause." But faced with the prospect of Uruguay's defection from Mercosur, he has toned down his language and sought to discourage the roadblocks, although the police have not intervened to halt them.

Though the two presidents have recently talked by phone about the standoff, they seem reluctant to make concessions that may offend their supporters. A Uruguayan congressman has suggested Vatican mediation, an idea that the papal nuncio quickly quashed. Argentina is talking about taking the case to the World Court in The Hague, where a decision would come only after the plants were operating.

In his most recent public declaration, Mr. Vázquez vowed that "construction of the plants will not be halted." As a way of criticizing the Argentines, he recalled the lyrics of an old tango, comparing their behavior to that of "the man who beats his wife because he fears she may cheat on him four or five years from now."

"That is exactly what is happening to us right now," he said. "They are inflicting real damage on us out of fear of some hypothetical damage we might cause them in the future."

Copyright 2006 The New York Times Company. Reprinted with permission.

In Review: Gualeguaychú Journal
A Back-Fence Dispute Crosses an International Border

Testing Your Comprehension

1. Why are the residents of Gualeguaychu blocking highways to Uruguay?
2. What are Jose Pouler's primary concerns about the construction of the paper mills?
3. How much money will be invested in the construction of the paper mills, and how much cellulose are they expected to produce for export?
4. Residents of the town "expressed frustration that their complaints were being ignored in both capitals." What are the two sources of their frustration?
5. What do the paper factories say they will do to protect the river?
6. What are the concerns of the Uruguayan authorities regarding commerce and trade as a result of the roadblocks?
7. In Fray Bentos, Uruguay, sentiment is just as strong in favor of the projects. Why?
8. Why has the dispute become a matter of sovereignty and national pride among Uruguayans?
9. What has the left-leaning government of Uruguay done to make clear to others in the Mercosur trade group that they are dissatisfied with their secondary role and the conduct of its neighbors?
10. What step is Argentina considering to resolve the dispute?

Weighing the Issues

1. Beyond the roadblock, what other alternatives could the protestors pursue to prevent the construction of the paper mills?
2. Many residents of Fray Bentos, Uruguay, welcome the jobs that the paper mills will bring. If you were the leader of an "economically dead" town like Fray Bentos, where would you stand on the issue of the paper mills, and what steps would you take to ensure that the town benefits in the safest way possible?
3. Sandra Caballero of Fray Bentos states, "We have faith that our government will be able to control emissions and punish the companies if they do something wrong." Do you have the same kind of faith in your own government? Why or why not?
4. Faced with the prospect of Uruguay's defection from the Mercosur trade group, President Kirchner of Argentina has toned down his language against the paper mills and sought to discourage the roadblocks. What would you do if you were in his position?
5. Uruguay's president objects that Argentina is "inflicting real damage on us out of fear of some hypothetical damage we might cause them in the future." Do you think this objection is justified? Why or why not?

More to the Story

Prepare a statement outlining the environmental and economic impacts of the paper mills as they will affect

- the town of Fray Bentos, Uruguay;
- the town of Gualeguaychu, Argentina;
- the countries of Uruguay and Argentina;
- the paper industry; and
- the global climate.

Use maps and data to support your points.

Useful Websites

http://www.travelpost.com/SA/Argentina/Entre_Rios/Gualeguaychu/map/4862665

http://www.alertnet.org/thenews/newsdesk/N30425323.htm

http://www.rodelu.com/uruguay/dpts/rionegroti-01.htm

http://americas.irc-online.org/am/3155

Recycled Inspection Reports and Other Water System Irregularities Stir Concerns Upstate

By Anthony DePalma
***The New York Times,* February 12, 2006**

They are the front line of New York City's vast upstate water system, responsible for delivering the one billion gallons a day of fresh, clean water that keeps the city alive.

But since 2001, these 873 inspectors, engineers and supervisors, along with their employer, the City Department of Environmental Protection, have been under court-ordered supervision for violating, of all things, environmental laws.

The department has had five years to retrain its water supply employees and change old habits. Citing improvements by the agency, a federal judge decided last week to end court supervision over the watershed in August.

The ruling was a relief for the department, but it upset local officials and upstate residents who say the department cannot be trusted, especially after a recent spate of incidents revealed continuing problems in the watershed.

Some of the incidents involved criminal activities, like theft; others appeared to stem from brazen negligence, like an inspector's repeated photocopying of the same weekly report on dam safety.

But taken together, the problems have heightened tensions in the 2,000-square-mile watershed. They have also worked to raise concerns that if relations between the city and the upstate communities deteriorate further, local officials may stop cooperating with the city, potentially costing New York billions of dollars.

"The D.E.P. has a culture of malfeasance and arrogance that's pervasive," said State Senator John J. Bonacic, a Republican whose district in Sullivan, Ulster and Delaware Counties includes six of the city's largest reservoirs.

Mr. Bonacic said he thought it might be too soon to end court oversight. He wants to keep the pressure on the department to clean up its act and become more responsive. It if does not, Mr. Bonacic said, he will urge the federal Environmental Protection Agency not to renew a crucial permit next year. Without that permit, the city could be forced to build a water filtration plant costing more than $8 billion.

Emily Lloyd, commissioner of the environmental department, said she was aware that upstate communities have concerns about the recent incidents and are worried about flooding. It is natural, she said, for such issues to surface during the process of renewing the filtration avoidance permit.

"It's their chance to get us to pay more attention to things they want us to respond to," Ms. Lloyd said.

Conveying clean water to New York City was the priority, not complying with detailed federal regulations

High on the list of concerns is the way the department takes care of its 22 dams, especially the one located where the city's water supply is farthest from the city. Some people who live in Schoharie County—much closer to the State House in Albany than to City Hall in Manhattan—panicked last November when they were told that the 78-year-old Gilboa Dam on the Schoharie Reservoir no longer meets state safety standards and needs emergency repairs. When frightened local residents pressed city officials, they discovered that because of neglected maintenance, valves that might have been used to lower the reservoir's water level were buried under 15 feet of silt and could not be opened.

Then, last month, a newspaper, The Times Herald-Record of Middletown, found that 70 percent of the weekly inspection reports for dams on two reservoirs in Sullivan County—the Neversink and the Roundout—were actually photocopies of earlier reports, with only the date and the weather conditions changed. City officials said the inspector in question had already been ordered to stop using photocopies. They asserted that the dams were safe and that the weekly reports were for internal use only.

But that was little comfort to critics.

"Dam safety is like riding in a Brink's truck—you can never let your guard down," said Howard R. Bartholomew, a resident of Middleburgh, in Schoharie County, who has been active in organizing local efforts to monitor the Gilboa Dam.

Mr. Bartholomew said he was distressed to learn that court supervision of the department's upstate operations would end soon. "I hate to see the judge reduce the pressure on them," Mr. Bartholomew said.

Department officials concede that it is difficult to supervise employees who work as far as 125 miles from the city.

"There's no getting around the fact that this was a very different culture," Ms. Lloyd said in an interview. "But it's not just culture. It's also history and habit."

In part, she said, what occurs now is the legacy of the department's predecessor, an autonomous agency created in 1905—long before the advent of environmental laws—to build the upstate reservoirs and pipelines. Conveying clean water to New York City

Photograph by Stewart Cairns for The New York Times
At the Gilboa Dam in Schoharie County, shown last November, valves that can lower the level of the reservoir were buried under 15 feet of silt.

was the priority, not complying with detailed federal regulations.

Although there have been improvements, the commissioner conceded that there was still work to be done to improve management of the watershed and the rest of the department.

Last September, the department's director of watershed programs, James D. Benson, was arrested, along with one of his associates, Lamberto S. Santos, and charged with possessing about $10,000 in city property, including generators, snowblowers and computers.

And last November, a former city employee, Dieter Greenfeld, was sentenced to two years' probation after he pleaded guilty to falsifying water quality records at what is known as the Catskill Lower Effluent Chamber in Westchester County when he worked there.

If that sort of thing could take place while the department was on probation, local officials wonder, what will happen when court supervision ends as scheduled in August?

When Judge Charles L. Brieant of the United States District Court in White Plains decided to end the watershed supervision, he did not terminate the court's oversight entirely.

Rather, he shifted it to the city, and the workers who operate 14 huge wastewater treatment plants.

In court last week, Ms. Lloyd acknowledged that workers had failed to repair backup generators at two wastewater treatment plants, causing millions of gallons of raw sewage to pour into New York waterways during the blackout of 2003.

The violations at the sewage plants, Anne C. Ryan, an assistant United States attorney, wrote in a report to the federal court, reflected "a mindset that persists in some components of D.E.P., namely, a failure to grasp that compliance is everyone's job."

"They suggest, at a minimum, that there is more work to be done to educate all D.E.P. employees about the requirements of environmental laws and the importance of complying with them," she wrote.

Copyright 2006 The New York Times Company. Reprinted with permission.

In Review: Recycled Inspection Reports and Other Water System Irregularities Stir Concerns Upstate

Testing Your Comprehension

1. Describe the federal judge's decision regarding supervision over New York City's watershed.
2. Problems within the City Department of Environmental Protection have heightened tensions in the watershed and raised concerns. What will happen if relations between the city and upstate communities deteriorate further?
3. What will Senator John Bonacic do if the department does not clean up its act and become more responsive, and what impact will this action have on New York City?
4. What is the primary safety issue that upstate residents are most concerned about?
5. What did concerned residents of Schoharie County discover after they were told that the Gilboa Dam no longer meets state safety standards and needs emergency repairs?
6. What violations were found in Sullivan County?
7. What is Emily Lloyd's description of the culture, history, and habit of the Department of Environmental Protection as it relates to the violations?
8. Describe two other violations that allegedly occurred on the part of the department's director of watershed programs and a former city employee.
9. Rather than terminate the court's oversight entirely, what did Judge Charles L. Brieant do?

10. What did an assistant U.S. attorney write in a report regarding the mindset in some components of D.E.P.?

Weighing the Issues

1. Do you agree with the judge's decision to end court oversight over the watershed as of August? Why or why not?
2. Officials from New York City, which receives the water from the watershed, manage this upstate New York watershed. What are the pros and cons of this management approach, which involves the supervision of employees who work as far as 125 miles from the city?
3. Put yourself in the position of a consultant to Emily Lloyd, commissioner of the environmental department. What steps would you advise her to take in revamping the department to become more effective stewards of the watershed?
4. What was Judge Brieant's rationale in shifting oversight responsibility for the watershed to the city (and the workers, who operate 14 huge wastewater treatment plants)? Will it be effective?
5. Anne C. Ryan, an assistant U.S. attorney, believes that "at a minimum . . . there is more work to be done to educate all D.E.P. employees about the requirements of environmental laws and the importance of complying with them." What are the best approaches to ensuring that this education occurs effectively?

More to the Story

Review the watershed protection and dam safety plans for the New York City watershed, and then familiarize yourself with the operations (including staffing structure) of the New York Department of Environmental Protection in the areas of water supply and protection. You are a consultant to the commissioner. Provide her with a memo describing how you would propose to reorganize the water programs' operations and structure to ensure a safer dam management program along with the conveyance of clean water.

Useful Websites

http://www.nyc.gov/html/dep/html/bureaus.html

http://www.gilboadaminfo.com/cgi-bin/teemz/teemz.cgi

http://www.thedailystar.com/news/stories/2005/11/04/damm2.html

http://www.watpa.org/lwv/water.html

Eating Well
Advisories on Fish and the Pitfalls of Good Intent

By Marian Burros
***The New York Times*, February 15, 2006**

Correction Appended

Shopping for fish these days is fraught with confusion. There is so much contradictory information about what is safe and what isn't. Some nutritionists are worried that people will throw up their hands and choose steak instead.

Part of the confusion is the result of a continuing public war between those scientists who think it is important to eat tuna and farmed salmon for their omega-3 fatty acids, despite the contaminants they contain, and those who think consumers should consider contaminants when deciding which fish to eat.

One contaminant, methylmercury, which can damage the nervous system and the brain in fetuses, infants and young children, is found in tuna, particularly albacore, or white meat. PCB's and dioxin, probable human carcinogens, are found in farmed salmon. But omega-3s, important nutrients in both types of fish, can prevent sudden heart attacks.

While the advice about tuna and salmon is directed at specific groups, there are studies that suggest heavy fish eaters could be at risk because mercury may contribute to cardiovascular disease, neurological problems and immune system problems. And even if the levels of PCB's and dioxin in farmed salmon are not high, nobody knows the cumulative effect of potential cancer-causing agents in the diet.

Among those who want people to eat tuna, no matter what its mercury content, are those who process it and put it in cans. Sometimes they say so directly through the United States Tuna Foundation, and sometimes they pay others, like the Center for Consumer Freedom, to say it for them.

In a news release from the latter organization, which is underwritten by tobacco, alcohol and restaurant interests, David Martosko, its director of research, says: "Americans need to be reminded that the health benefits of eating fish are very real, while the risks are imaginary."

Mr. Martosko's statement is counter to the warnings issued in 2004 by the Environmental Protection Agency and the Food and Drug Administration about the hazards of mercury for women of childbearing age and young children. The agencies said those groups should eat no more than six ounces of albacore tuna a week, but could safely eat up to 12 ounces per week of low-mercury fish like shrimp, canned light tuna, salmon, pollock and catfish. At the same time the agencies continued to recommend that those groups limit their intake of shark, swordfish, king mackerel and tilefish.

Photographs by Andrew Scrivani for The New York Times

THE REPLACEMENTS Salmon replaces tuna in an old-fashioned sandwich, and is featured in a spread made with yogurt, Neufchatel cream cheese and pecans, next page.

Concerns about farmed salmon are more recent, but scientists who have studied the pollutants they contain strongly suggest that the same vulnerable population, which includes pregnant women, should choose fresh, frozen or canned wild salmon, which is comparatively uncontaminated with PCB's and dioxin.

Some nutritionists are concerned that such warnings confuse people and may make them reduce their consumption of all tuna and salmon. Wittingly or unwittingly, these nutritionists find themselves on the same side as the tuna industry.

Rebecca Goldberg, a senior scientist for Environmental Defense, an advocacy group often at odds with the food and agriculture industries, finds the nutritionists' response wrongheaded. "I just get disgusted with the view that the public is so dull-headed that they can't understand that fish are generally good for you, but some kinds should be avoided because they are heavily contaminated," she said.

If fish sales are any guide, many people appear to understand that fish is good for them but that tuna should be eaten sparingly. Sales of canned tuna from October 2004 to October 2005 dropped 9.8 percent, according to Information Resources Inc., a market research firm. But fish consumption has increased 12 percent since 2001, up from 14.8 pounds per person a year to 16.6 pounds per person in 2004.

Tuna canners—Bumble Bee, StarKist and Chicken of the Sea—through the United States Tuna Foundation, are spending significant sums to counter the decline, through full-page advertisements and the foundation's Web site, tunafacts.com, and by fighting the California attorney general's lawsuit against the canners for not warning consumers about the mercury in their tuna.

Some of their efforts, however, are less obviously related to them. The Center for Consumer Freedom, for example, has a Web site, fishscam.com, that says the risks from mercury are theoretical. The tuna foundation gave $45,000 to the University of Maryland's newly formed Center for Food, Nutrition and Agriculture Policy to create the Web site realmercuryfacts.org. Public relations for that site are handled by Ruder Finn, whose client is the tuna foundation.

In addition, most of the $500,000 paid for a scientific study of the risks and benefits of hypothetical changes in fish consumption, conducted by the Harvard Center for Risk Analysis, was paid for by the United States Tuna Foundation, but the foundation is not listed as a funder. Funders listed are the National Food Processors Association Research Foundation, a trade association now known as the Food Products Association, and the Fisheries Scholarship Fund, part of the National Fisheries Institute, a seafood industry trade association.

Critics say the failure to identify the source of the funding is a conflict of interest. Edward Groth, an environmental health expert retired from Consumers Union, said: "No matter how well they did their analysis, since an affected industry paid for it, its credibility is suspect."

Joshua T. Cohen, the study's lead researcher, now a research associate at the Institute of Clinical Research and Health Policy Studies at Tufts-New England Medical Center, said he saw nothing wrong with the omission of the primary funding source.

"No one is hiding anything," he said. "It never occurred to me anyone would think National Food Processors Association was less industry than Bumble Bee tuna."

The study, in the American Journal of Preventive Medicine, used hypothetical situations based on how people might react to dietary advice. Among its conclusions: if pregnant women ate low-mercury fish without decreasing total fish consumption there would be public health benefits, but if all adults ate less fish there would be a large negative impact on cardiovascular health.

The latter conclusion received all the attention, abetted by the headline on a university news release: "Study Finds Government Advisories on Fish Consumption and Mercury May Do More Harm Than Good." The tuna industry is making good use of that headline.

If fish sales are any guide, many people appear to understand that fish is good for them but that tuna should be eaten sparingly

The journal also published an editorial by Walter Willett, professor of epidemiology and nutrition at the Harvard School of Public Health. He, too, questioned the unintended consequences of the government's advice to reduce consumption of fish high in mercury.

He also criticized a 2004 study, published in Science, that reported that farmed salmon had high levels of PCB's and dioxin.

"That publication was particularly troublesome, perhaps even irresponsible," he wrote, "because the implied health consequences were based on hypothetical calculations and very small lifetime risks." Despite government statistics to the contrary, the report, he said, "almost certainly contributed to a reduction in fish consumption that "likely caused substantial numbers of premature deaths."

Dr. Willett said in a telephone interview that he stood by his editorial.

"I thought it needed to be said because people are confused," he said.

Dr. Willett has received intense criticism. A letter to be published in the journal from the authors of the Science study calls his comments defamatory, inaccurate and scurrilous. Marion Nestle, a professor of nutrition at New York University, said his statement was "astonishing."

But the public is really not faced with a Hobson's choice. It can always get plenty of omega-3's from canned wild salmon, cheap and available year-round and low in contaminants.

Correction: February 22, 2006, Wednesday The Eating Well column last Wednesday, about tuna, misidentified the plaintiff in a lawsuit accusing canners of not warning consumers about the mercury in their products. The original plaintiff was the Public Media Center, a nonprofit media and consumer advocacy agency in San Francisco; the California attorney general joined the lawsuit later.

Copyright 2006 The New York Times Company. Reprinted with permission.

In Review: Eating Well
Advisories on Fish and the Pitfalls of Good Intent

Testing Your Comprehension

1. The article describes the confusion surrounding shopping for fish as the result of a "continuing public war." Who is involved in the "war," and what do they believe?
2. Name three contaminants that can be found in seafood.
3. Why would heavy fish eaters be at risk from contaminants?
4. How do tuna processors and canners communicate the message that people should eat tuna?
5. What did the Environmental Protection Agency and the Food and Drug Administration say in 2004 regarding the hazards of mercury for women of childbearing age and young children?
6. What do scientists who have studied farmed salmon recommend to these populations regarding their consumption of salmon, and why?
7. Some nutritionists are concerned about these warnings. Why?
8. How did sales of canned tuna change from October 2004 to October 2005?
9. What entities paid for the scientific study of the risks and benefits of hypothetical changes in fish consumption, conducted by the Harvard Center for Risk Analysis?
10. What were the conclusions of the study?

Weighing the Issues

1. In a news release from the Center for Consumer Freedom, David Martosko says, "Americans need to be reminded that the health benefits of eating fish are very real, while the risks are imaginary." Do you agree? Why or why not?
2. Will you alter your consumption of fish in any way as a result of this article? Why or why not? If yes, how will your consumption change?
3. Edward Groth, formerly of the Consumers Union, said: "No matter how well [the Harvard Center for Risk Analysis] did their analysis, since an affected industry paid for it, its credibility is suspect." In many cases, private foundations pay for studies produced by non-profit organizations. Would this affect the credibility of a study in the same way as an industry-funded study? Explain your answer.
4. What do you believe are the most credible sources of information on a topic such as food safety?
5. Does this article present a point of view in support of one party or another in the tuna debate? If so, explain how that point of view is communicated.

More to the Story

1. You are the head of a non-profit organization working to educate the public on the dangers of contaminated

seafood. What strategies would you employ to get your message across to various audiences? List your top five audiences, a message for each, and a one-sentence strategy describing how you will communicate that message to your audience.

2. You are the head of a non-profit organization of salmon farmers who produce a large amount of fish (both farmed and wild) for global consumption. You want to educate the public and others on the benefits of eating fish high in omega-3 fatty acids. What strategies would you employ to get your message across to various audiences? List your top five audiences, a message for each, and a one-sentence strategy describing how you will communicate that message to your audience.

Useful Websites

http://www.mbayaq.org/cr/seafoodwatch.asp

http://www.sciencemag.org/cgi/content/abstract/303/5655/226

http://www.pubmedcentral.nih.gov/articlerender.fcgi?artid=1257546

http://www.salmonoftheamericas.com/

Out of Old Mines' Muck Rises New Reclamation Model for West

By Kirk Johnson
***The New York Times,* March 4, 2006**

BRECKENRIDGE, Colo.—Pollution saves fish is like man bites dog.

But that is what happened in this mining town turned resort as contaminated water from abandoned gold mines drained into French Gulch creek beginning after World War II. The pollution created a barrier that isolated the native Colorado cutthroat trout—a threatened species—from nonnative competitors like the rainbow trout that could not fight their way through the muck. The cutthroats thrived in their accidental sanctuary above the mines, where the water was clean.

Now Brian Lorch has to worry about keeping it that way.

"If we're successful in cleaning up French Gulch, we could make it possible for the competitors to move up," said Mr. Lorch, an open space and trails resource specialist for Summit County.

Fish are only one of many wrinkles on a storied piece of mining land called the Golden Horseshoe. In what environmental regulators and academics say was the first deal of its kind in the nation, Breckenridge and Summit County last fall bought 1,840 acres of the 8,000-acre Horseshoe gold field—most of the rest is already national forest—from a company that had obtained the assets when the mines went bankrupt.

With ownership came responsibility for the environmental problems and the management of a landscape that created Breckenridge in the days of the Gold Rush, but that also suffered some of the worst of mining's effects.

"It's the first time in the nation where the E.P.A. entered into an agreement with a purchaser of a site like this who agreed to do a cleanup," said Andrea Madigan, a lawyer for the Environmental Protection Agency. Usually, she said, cleanups involve dragooning responsible parties and compelling their actions. Volunteers do not usually line up.

Photographs by Kevin Moloney for The New York Times

Kristin Frederickson, left, a Harvard graduate student, with Scott Reid, a town planner, inspecting an old mining mill in Breckenridge, Colo.

Local officials admit that the risks are substantial. Just digging a hole, if it uncovers or disturbs environmental problems, could expose the new owners to liabilities as if they had caused the pollution itself. The buyers agreed to spend about $2 million for environmental cleanup, including a $1.2 million water-treatment plant that would have to be operated in perpetuity at a cost of about $90,000 a year, on top of the $8.2 million purchase price.

But not acting, local officials said, would have been even riskier. In years past, they said, the property was shielded from development pressures by concerns about pollution and cleanup costs, much the way the upstream cutthroat trout were isolated.

Explosive growth in the Colorado Rockies over the last decade began to crack those barriers, and town planners said they feared that the property's former owners would begin selling off less damaged parcels to pay for the cleanup and water-treatment that the E.P.A. wanted.

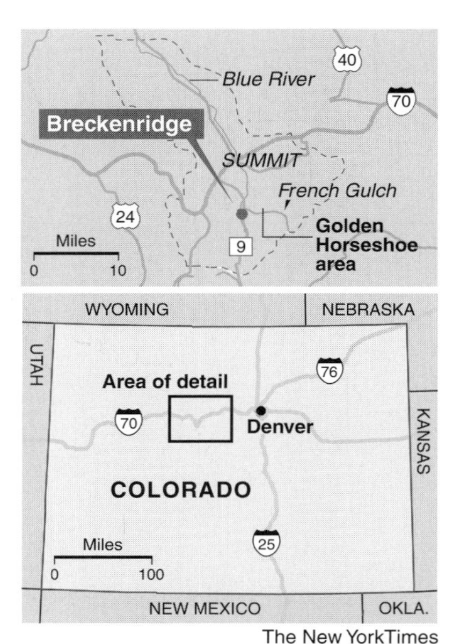

The New YorkTimes

A project to transform an old mining area into a vibrant resort

And if the Horseshoe was developed for housing, they said, then the roads into it would open up large tracts of backcountry that Breckenridge wants

Claire Agre, left, a Harvard graduate student, recently presented ideas for developing the old mining area.

preserved for its outdoor recreation economy.

The Horseshoe's pollutants, mainly heavy metals like cadmium and zinc, are ecologically harmful, but are not considered a human health risk by the E.P.A.

"We needed to get the problem sites in order to protect the whole," said Todd Robertson, the county's director of open space and trails.

Academics and experts on mine reclamation—one of the biggest environmental problems of the West, where there are perhaps 500,000 abandoned mines—say that Breckenridge's groundbreaking path could change how mine reclamation works. With ownership of the pollution and control of the land, they say, comes the power to shape the post-mining landscape in a way that goes far beyond just cleaning it up.

"Breckenridge can lead the way," said Alan Berger, an associate professor of landscape architecture at Harvard and founder of the Project for Reclamation Excellence, a group at the design school that works on reclaiming land damaged by resource extraction. "The opportunities of what the town and county can do here are completely open-ended."

"It's the first time in the nation where the E.P.A. entered into an agreement with a purchaser of a site like this who agreed to do a cleanup."

And so are the burdens. The property is hatch-marked by miles of unmarked and unmapped trails carved by generations of backcountry users at a time when no owner was around to say boo. The new owners are bracing for what they expect will be contentious public meetings beginning this spring as managers decide which trails to keep open and who may use them. The town favors things like hiking and biking, while the county wants to make sure that motorized users have their say as well.

Mining's legacy on the forest is another headache. In gold's heyday, lumber was needed for mills and tunnels, and by the late 1800's the Horseshoe was stripped. The result today is a narrow monoculture in which the oldest trees are about 120 years old—mostly lodgepole pines.

Pine beetles, which have ravaged vast parts of the West, are just hitting this part of Colorado. Local officials warn that the bugs, which love mature lodgepoles, could kill 80 percent of Horseshoe's trees.

But Breckenridge, which has carved a tourist niche around its mining history and historic buildings also wants to incorporate the story of the Horseshoe into the fabric of the local economy. That means thinking about mines and miners, and how they gave rise to Breckenridge.

That is where Claire Agre, a Harvard graduate student in landscape architecture, enters the picture.

Ms. Agre, with eight other graduate students from Harvard's Graduate School of Design who work with Professor Berger, came to Breckenridge in February with their professor to think big thoughts about what the Golden Horseshoe might become.

Her initial idea was that history, pollution, tourism and recreation all fit together perfectly. "The history of mining is part of the town's identity—I think the reclamation itself could become part of that identity," she said. Perhaps, she mused, a cultural trail could be carved through the property that would include the abandoned mines and an interactive display about the impact of mining and the technology of the cleanup. The students will present their ideas to the town and county in May.

The cutthroat trout, meanwhile, which get their name from the little filigree of red on their gills, are apparently safe for now.

Mr. Lorch said state wildlife officials had looked at two road crossings under which French Gulch passes in its journey from the Horseshoe and concluded that the artificial waterfalls from the pipes under those roads provided adequate barriers to downstream interlopers, clean water or no. Signs will go up soon, he said, warning that wildlife officials must be notified before any changes can be made to the crucial crossings.

Copyright 2006 The New York Times Company. Reprinted with permission.

In Review: Out of Old Mines' Muck Rises New Reclamation Model for West

Testing Your Comprehension

1. What did the water pollution from the abandoned gold mines in French Gulch creek do to protect the native Colorado cutthroat trout from nonnative competitors?
2. What is Brian Lorch's concern about what could happen if French Gulch is cleaned up?
3. Describe the deal that occurred that could make it possible for French Gulch to be cleaned up.
4. What makes this deal the first of its kind in the nation?
5. What did the buyers agree to in terms of cleanup, and how much will it cost?
6. Why is it rare for volunteers to purchase contaminated land?
7. What were the town planners' fears about what could happen if a buyer did not step in to purchase the Golden Horseshoe?
8. As town managers prepare to decide which trails to keep open and who may use them, what activities does the town appear to favor in comparison to the county?
9. Name two problems affecting the forest itself.
10. Wildlife officials have concluded that the cutthroat trout are safe for now. Why?

Weighing the Issues

1. Brian Lorch, open space and trails specialist for Summit County, is concerned about the effect that cleaning up French Gulch will have on the cutthroat trout population. Can you think of other situations in which the new management of the Golden Horseshoe might unwittingly present ecological threats?
2. Can you envision this type of land deal happening elsewhere? If so, in what types of places would it occur?
3. Academics and reclamation experts believe that with ownership of the pollution and control of the land comes the power to "shape the post-mining landscape in a way that goes far beyond just cleaning it up." What do they mean by this? Provide examples.
4. The town and county have somewhat opposing views on which trails to keep open and who may use them. If you were a mediator in this debate, how would you manage the discussion?
5. You are a member of the team from the Harvard Graduate School of Design charged with developing ideas how to incorporate the story of the Horseshoe into the fabric of the local economy. Describe your own proposal in the same level of detail that Claire Agre described hers.

More to the Story

The French Gulch project is a model for mine reclamation that will restore almost 2,000 acres of the 8,000-acre Horseshoe gold field. Among other benefits, town planners expect to keep open miles of back-country trails for recreational use, which will provide a major amenity to Breckenridge and Summit Counties.

At the same time, students from the Harvard Graduate School of Design have proposed alternative scenarios designed to reincorporate the mining culture of Breckenridge back into the town's history. Based on what you can learn about the Breckenridge area and the students' proposals, which approaches do you think will be most effective in preserving the town's culture while also meeting the current needs of retirees, weekend tourists, and other visitors and residents to the town? What strategies and features might you avoid, and which would you emphasize?

Useful Websites

http://projects.gsd.harvard.edu/prex/symposia.htm

http://www.redlodgeclearinghouse.org/stories/frenchgulch.html

http://www.abch2o.com/welloro1.htm

http://www.epa.gov/region8/superfund/co/frenchgulch/

http://www.poplarhouse.com/

New Uses for Glut of Small Logs From Thinning of Forests

By Jim Robbins
***The New York Times,* January 10, 2006**

DARBY, Mont.—Five years ago, intense forest fires around this logging and tourist town burned more than 350,000 acres of forest. Today huge swaths of charred trees cover the mountainsides.

Partly in response to these fires and others on national forest land elsewhere in the West, President Bush introduced the Healthy Forest Initiative in 2002 to reduce the wildfire threat to towns surrounded by publicly owned forests. As work crews thin stands of trees, as called for in the initiative, one result has been a glut of logs smaller than eight inches in diameter.

Until recently, most small trees were collected in piles and burned, but now businesses and the Forest Service have begun looking for uses for the tiny trees.

"It's high cost, low value and a lot of pieces to handle, which takes time and effort," said Dave Atkins, head of the Forest Service's Fuels for Schools program for several Western states.

Although loggers might receive $90 a ton for house logs, Mr. Atkins said, they are paid less than half that for smaller trees.

Slowly, however, the small-diameter movement, helped along by federal grants and Forest Service research, is helping to find new uses for smaller trees, like heating schools and hospitals and construction materials, including particle board, flooring and laminated beams.

Peter Stark of Missoula, a freelance writer, wanted to thin his 80 acres of forest clogged with downed timber and crowded trees to prevent a fire but could not afford to do it, since clearing usually costs $300 to $1,000 an acre.

He eventually found someone to remove the trees, most six or seven inches across, and the money he was paid for them covered the cost of thinning.

Photographs by Jim Robbins for The New York Times

JUST A TRIM

Peter Stark last month inspecting 80 acres of his forest that he had thinned for fire prevention. The effort led him to investigate uses for small diameter trees, and to start a new company to make wood products from them.

At the same time, he was building a dance floor for his wife, Amy Ragsdale, who teaches dance at the University of Montana. Shocked at the cost of hardwood, Mr. Stark realized that he might be able to turn the waste trees into flooring.

Mr. Stark bought back 25 tons of the larch trees and found a custom sawmill that could handle small diameters to

THINKING BIG
Wood from small trees was used in the construction of the library in Darby, Mont.

turn them into tongue-and-groove flooring. The floor turned out so well that Mr. Stark formed a company, North Slope Sustainable Wood, with two partners, to market small diameter larch, the hardest of the soft woods, from forests being thinned.

He sees such activity as a solution to the controversy over logging in Western forests.

"I'm a tree hugger," he said. "If we can take the small trees and leave the big ones, the loggers and environmentalists are both happy."

Significant numbers of Westerners see small trees as the future of the timber industry, simply because there are so few big trees left.

"Significant numbers of Westerners see small trees as the future of the timber"

"Years ago, we utilized logs that were mostly over 50 inches in diameter," said Gordy Sanders, resource manager for Pyramid Lumber in Seeley Lake, which has retooled to use small-diameter timber. "Now, if we see one of those a year we're amazed."

Another project, at the Forest Service's laboratory at the State and Private Technology Marketing Unit in Madison, Wis., used small-diameter trees in a new library here, in the town that bore the brunt of the fires.

"This library was a response to the fires," Veryl Kosteczko, chairwoman of the library board, said as she pointed out the roof beams that are all six inches or so in diameter. "We utilized underutilized wood that used to be left as trash."

Another use of small logs is as biomass to be turned into fuel. Under its Fuels for Schools program, the Forest Service is giving grants up to $400,000 for schools and other public buildings to build furnaces that burn biomass.

The three public schools in Darby are heated by a large $800,000 furnace that burns a steady stream of tiny branches and wood chips arriving by conveyor. Rick Scheele, the maintenance supervisor for the schools and the mayor of Darby, estimates that heating the school with diesel this year would have cost $125,000 and that using biomass will cost $28,000.

"It's allowed a few extra teachers to stick around," Mr. Scheele said. "It's been pretty tight around here."

For the moment, environmentalists are watching the small-diameter movement warily.

"We support hazardous-fuels reduction," said Bob Ekey, Northern Rockies regional director for the Wilderness Society. "But we want to make sure it's done well, and done right, so we don't create more demand than the land can sustain."

Copyright 2006 The New York Times Company. Reprinted with permission.

In Review: New Uses for Glut of Small Logs From Thinning of Forests

Testing Your Comprehension

1. What happened to the forest surrounding Darby, Montana five years ago?
2. What is the purpose of the Healthy Forest Initiative?
3. In terms of timber supply, what has been the result of recent efforts to thin stands of trees as called for in the Health Forest Initiative?
4. Name the challenges associated with finding uses for these tiny trees.

5. How much money do loggers receive for house logs as compared to the smaller trees?
6. Name two uses for the smaller trees.
7. Why do significant numbers of Westerners see small trees as the future of the timber industry?
8. What is the goal of the Fuels for Schools program?
9. How much would heating the Darby school with diesel have cost this year as compared to using biomass?
10. Describe the perspective of the Wilderness Society representative regarding the small-diameter movement.

Weighing the Issues

1. What other uses can you identify for the small-diameter trees?
2. New uses of the small-diameter trees were driven more as a result of an abundance of small-diameter wood (supply) than of demand. What steps could be taken to increase demand for the wood?
3. Bob Ekey of the Wilderness Society is concerned that the small-diameter movement may create an unsustainable demand. What are the potential risks of that possible outcome?
4. What steps should environmentalists take to ensure that the small-diameter movement does not become unsustainable?
5. The glut of small logs resulted from increased thinning of trees to reduce the wildfire threat to towns surrounded by publicly owned forests. Beyond thinning, what other approaches might be effective in reducing the wildfire threat to towns?

More to the Story

You are the director of a regional office of a major conservation group based in the Northwest, and you are concerned about the possibility that new demands for small-diameter lumber will lead to unsustainable harvests of the landscape. Formulate a plan to prevent this from occurring. If you plan to collaborate with other stakeholders, please describe these stakeholders, their positions on the issue, how you plan to engage them, your strategies and tactics, and your preferred outcome.

Useful Websites

http://www.whitehouse.gov/infocus/healthyforests/

http://www.greenleafforestry.com/greenleafabout_007.htm

http://www.fpl.fs.fed.us/documnts/pdf2001/levan01a.pdf

Canada to Shield 5 Million Forest Acres

By Clifford Krauss
***The New York Times,* February 7, 2006**

HARTLEY BAY, British Columbia, Feb. 4—In this sodden land of glacier-cut fjords and giant moss-draped cedars, a myth is told by the Gitga'at people to explain the presence of black bears with a rare recessive gene that makes them white as snow.

The Raven deity swooped down on the land at the end of an ice age and decided that one out of every 10 black bears born from that moment on would be bleached as "spirit bears." It was to be a reminder to future generations that the world must be kept pristine.

On Tuesday, an improbable assemblage of officials from the provincial government, coastal Native Canadian nations, logging companies and environmental groups will announce an agreement that they say will accomplish that mission in the home of the spirit bear, an area that is also the world's largest remaining intact temperate coastal rain forest.

"The customer doesn't want products with protestors chained to it"

A wilderness of close to five million acres, almost the size of New Jersey, in what is commonly called the Great Bear Rain Forest or the Amazon of the North will be kept off limits to loggers in an agreement that the disparate parties describe as a crossroads in their relations.

The agreement comes after more than a decade of talks, international boycott campaigns against Great Bear wood products and sit-ins in the forests by Native Canadians and environmentalists, who chained themselves to logging equipment.

Photographs by Bayne Stanley for The New York Times

Princess Royal Island, in British Columbia, is part of the Great Bear Rain Forest, the largest remaining temperate coastal rain forest. The area, more than 15 million acres, has only about 25,000 residents.

The process has already inspired similar efforts to save the Canadian boreal forest, to the north, and suggestions that the agreement could be a model for preservation in the Amazon and other threatened forests.

Scientists say the agreement should preserve not only the few hundred spirit bears and other black bears, but also one of the highest concentrations of grizzly bears in North America as well as unique subspecies of goshawks, coastal wolves, Sitka blacktail deer and mountain goats.

"It's like a revolution," said Merran Smith, director of the British Columbia Coastal Program of Forest Ethics, an environmental group. "It's a new way of thinking about how you do forestry. It's about approaching business with a conservation motive up front, instead of an industrial approach to the forest."

Under the agreement, the loggers will be guaranteed a right to work in 10 million acres of the forest, which some environmentalists criticize. But they will be obliged to cut selectively: away from critical watersheds, bear dens and fish spawning grounds, negotiators said.

"There's a new era dawning in British Columbia," said Gordon Campbell, the province's premier. "You have to establish what you value, and work together. This collaboration is something we have to take into the future, and it is something the world can learn from."

As a sign of new Native power gained in recent court cases, many areas that will be preserved or selectively logged have been chosen based on the oral tradition of Native groups and the opinions of their elders. These include areas with cultural significance like ancient cemeteries, or those with medicinal herbs and cedars big enough to make totem poles, canoes and long houses.

If the federal government agrees, more than $100 million will also be raised by governments and foundations to start ecotourism lodges, shellfish aquaculture and other environmentally sustainable economic activities for the 25,000 people who live in the region.

"Now we can manage our destiny," said Ross Wilson, chairman of the tribal council of the Heiltsuk, one of the Native nations involved. "Without this agreement, we would be going to court forever and we would have to put our children and old ladies dressed in button blankets in the way of the chainsaws," he added, referring to the ceremonial dress worn in past protests.

Among the supporters of the agreement are some of the biggest players in Canadian lumber and paper, including Western Forest Products, Interfor and Canfor.

"It's a cultural shift," said Shawn Kenmuir, an area manager for Triumph Timber, which has already forsaken old clear-cut practices and begun consulting with the Gitga'at before cutting on their traditional lands. "We've started the transition from entitlement to collaboration."

The forest represents a quarter of what remains of coastal temperate rain forests in the world.

Because 15 feet of rain can fall in a year, the Great Bear has never suffered a major forest fire. That has allowed some of the tallest and oldest trees on earth to thrive, including cedars more than a thousand years old.

Shawn Kenmuir, area manager for Triumph Timber, which already avoids clear-cut logging, a practice that has sometimes led to boycotts.

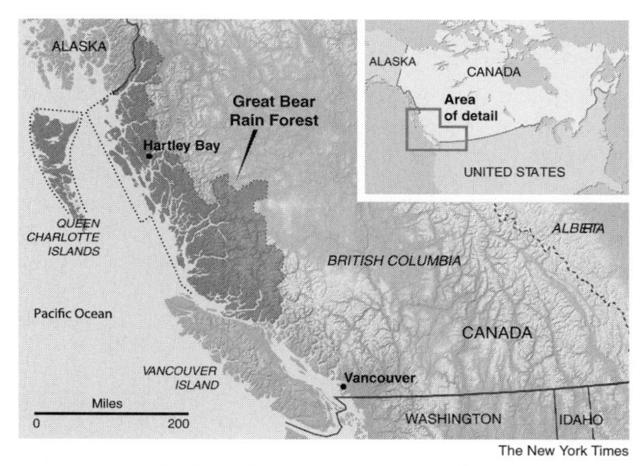

The New York Times

Sacred sites in the Great Bear Rain Forest are to be protected.

An estimated 20 percent of the world's remaining wild salmon swim through the forest's fjords, including coho and sockeye, whose spawning grounds were threatened by erosion caused by past logging. Largely intact because of its remoteness, the forest contains an abundance of wolverines, bats, peregrine falcons, marbled murrelet sea birds and coastal tailed frogs.

The ecological richness is immediately apparent to the few people who visit. Within minutes of a recent helicopter visit to Princess Royal Island, in the heart of the rain forest, a group of visitors saw a pack of six gray and black wolves, a seal and numerous bald eagles and swans.

"Look at the forest move," said Marven Robinson, 36, a Gitga'at guide, as eagles glided through the moist air and the wolf pack played hide-and-seek with the visitors along a channel of diaphanous water. "As long as there is a spirit bear, we're going the right way."

The efforts to save the rain forest began a decade ago, as lumber companies that had already cut most of the old-growth forest around British Columbia, by far Canada's richest forestry province, began moving into the Great Bear.

A deluge of postcards and demonstrations by groups like the Sierra Club and Greenpeace at shareholders meetings and retail outlets pressed American, Japanese and European hardware chains to shun products from the area.

By 1999, when the Home Depot announced it would phase out sales of wood from the Great Bear and other endangered old forests, some lumber companies were shifting their approach, agreeing to work with the environmentalists.

MacMillan Bloedel, before it was acquired by Weyerhaeuser, broke ranks with the industry and promised in 1998 to phase out clear-cutting on the British Columbia coast. Other companies gradually fell into line.

"The customer doesn't want products with protesters chained to it," said Patrick Armstrong, a consultant who served as a negotiator for the lumber companies. "We're dealing with old-growth forests with charismatic wildlife."

Once Mr. Armstrong sat at the opposite side of the bargaining table from the environmentalists, but now he works closely with them. "This needs to be celebrated — it's a big, big deal," he said. "Everyone had a greater interest in resolving the problems than continuing the conflict."

Copyright 2006 The New York Times Company. Reprinted with permission.

In Review: Canada to Shield 5 Million Forest Acres

Testing Your Comprehension

1. What is the reminder of the "spirit bear"?
2. What are primary goals of the agreement for the Great Bear Rain Forest?
3. In addition to the spirit bears and black bears, what other species will be protected as a result of the agreement?
4. Merran Smith, of the British Columbia Coastal Program of Forest Ethics, describes the agreement as a "new way of thinking about how you do forestry." What does he mean by this?
5. Within the 10 million acres of forest where the loggers will be guaranteed a right to work, what conservation obligations will they be required to meet?
6. According to what types of criteria did the Native groups choose areas to be preserved or selectively logged?
7. What additional conservation activities are envisioned if the federal government agrees and an additional $100 million is raised?
8. What portion of the world's remaining coastal temperate rain forests does the Great Bear Rain Forest represent?
9. How much of the world's remaining wild salmon swims through the forest?
10. Name three actions that pressured lumber companies to shift their approach in the Great Bear Rain Forest.

Weighing the Issues

1. Under the agreement to protect the forest, loggers will be obliged to selectively choose their sites according to specific guidelines. What are the challenges associated with implementing this part of the agreement?
2. Think of another environmental conflict that involves a large degree of public protest. What would it take to bring the opposing parties to an agreement as was done with the Great Bear Rain Forest?
3. Merran Smith, director of the British Columbia Coastal Program of Forest Ethics, says the agreement is "like a revolution . . . . it's about approaching business with a conservation motive up front, instead of an industrial approach to the forest." In what ways will this conservation motive benefit the timber companies?
4. You are in charge of raising more than $100 million to start ecotourism lodges, shellfish aquaculture and other environmentally sustainable economic activities. How would you proceed?
5. The article states that the collaboration process that resulted in the agreement could serve as a model for preservation in the Amazon. Can you foresee any differences in how such an agreement might be pursued in the Amazon versus Canada?

More to the Story

Noting the timing of the Web-based tools available to you, trace the evolution of the Great Bear Rain Forest Campaign. What were the major turning points that led to its success? Are there any ways in which the resolution of the debate could have been expedited? Why or why not? Is there anything unique about the way government and NGOs operate in North America that might make this a one-of-a-kind success? Why or why not?

Useful Websites

http://www.savethegreatbear.org/

http://www.fanweb.org/gbr/index.html

http://www.greenpeace.org/international/news/great-bear-saved-123987

Competing Plans to Repair New Orleans Flood Protection

By John Schwartz
***The New York Times*, January 22, 2006**

At the halfway mark between the onslaught of Hurricane Katrina last year and the beginning of the 2006 hurricane season on June 1, the Army Corps of Engineers has completed only 16 percent of its planned repairs to New Orleans's battered flood protection system, according to corps representatives.

The corps says its work is on track for restoring the system to its pre-hurricane strength by the June 1 deadline, but in the meantime many groups that have studied the disaster are coming up with proposals of their own that they say could be cheaper, faster or stronger.

The Bring New Orleans Back Commission, the group formed by Mayor C. Ray Nagin to produce a blueprint for the city's recovery, issued a proposal on Wednesday to upgrade hurricane protection with measures beyond what the corps has called for. To prevent storm surges from pushing into the city's drainage canals, the commission proposed a series of jetties to stand in front of the three canals, which it says could be built quickly and cheaply and provide New Orleans with some much-needed peace of mind.

"There is, very much, a tension between things that can be done quickly versus those that might take a little longer," Lawrence Roth, deputy executive director of the American Society of Civil Engineers, said in a telephone interview on Friday. His group has weighed in with far-reaching recommendations, and other groups are preparing proposals of their own.

The mayor's commission also proposed a network of dams that would block or slow the opening between the Inner Harbor Navigational Canal and Lake Pontchartrain, and block storm surges from flowing up the Mississippi River Gulf Outlet, a navigation channel that has been blamed for a storm surge funnel effect that increased the damage to eastern New Orleans.

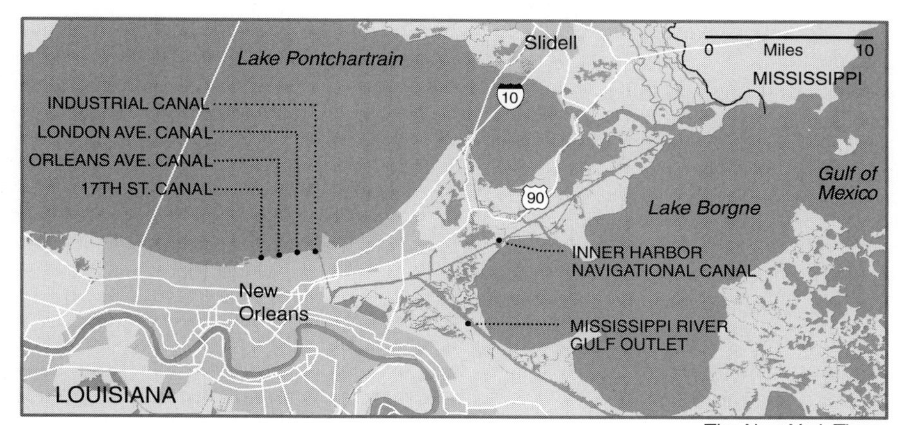

New Orleans's canals are a focus of flood-protection proposals.

The group is also calling for long-term flood-control structures that would block or slow surges at the two passes between Lake Pontchartrain and the Gulf of Mexico.

The fast-track structures would cost \$100 million to \$170 million, according to the commission's estimates, a fraction of the \$3.1 billion the federal government has proposed spending on flood control measures in the area. The commission said its proposals would not interfere with any of the corps's plans, but would be add-ons that complement the current plans.

The proposals have not yet found broad support among other engineering experts who have been working on strengthening New Orleans's storm defenses, but Dan Hitchings, the director of the corps's Task Force Hope, which is coordinating the hurricane response in Louisiana and Mississippi, said the plans were welcome and would be examined.

Mr. Roth, of the Society of Civil Engineers, said there would always be competing ideas about how to improve flood protection. The idea of jetties, he said, might be made moot by closing off the canals and putting in new pumping stations at the lake, as the corps has planned.

"Many different people can look at a problem and come up with many different solutions, all with tradeoffs," he said. "Which would be better—jetties or a pump station? You might never get an answer to that."

Meanwhile, the corps's work to restore flood protection to its pre-hurricane levels continues around the clock. This month the corps solicited bids for building temporary closures and pumps at the mouths of the city's three drainage canals, and it is rebuilding long stretches of levee in St. Bernard Parish and along the Inner Harbor Navigational Canal.

The corps is looking to measures that will further strengthen the flood protection system, including restoring levees to their originally designed heights. These measures can be in place by September 2007, according to the corps.

Beyond that, the corps has embarked on a two-year, \$8 million study to determine how to strengthen the hurricane protection system for New Orleans and southern Louisiana. A preliminary version of that report is due in June.

While the Bush administration's top official on Gulf Coast reconstruction,

Donald Powell, has said the government will build a system that is "better and stronger" than what was there before, the administration has not committed to what the people of New Orleans desperately want: protection from Category 5 storms, the toughest that nature can dish out.

Mr. Hitchings said that the corps was slightly behind schedule but that he expected things to move quickly. "It's not linear," he said, because the "gear-up time" to get contractors in place and to make materials like the enormous quantities of soil available was so great.

Now "they're really moving out," he said. The corps built 30 days of weather delays into the schedule, he said, and with a little help from favorable weather, "I'm very optimistic that they will regain their schedule and in the end get it all finished with plenty of time."

"There is . . . a tension between things that can be done quickly versus those that might take a little longer"

The corps's long-term study, he said, would probably have a lot in common with the outside proposals that are beginning to flow in, but "right now, we're focused on the very near term."

The engineering society is investigating the failure of the levees and is working with the groups that will monitor the corps's progress. Its recommendations include "armoring" the dry side of levees so they are not eroded away from underneath if water spills over the top. Without armoring, Mr. Roth said, "failure is catastrophic because it causes the wall to fail."

The corps has said that the armoring process, like other projects that would go beyond the restoration of the levees to pre-hurricane strength, will have to be approved by Congress.

"It's going to take people being willing to take a chance, to be bold, to sort out Louisiana's levee problems," said Ivor van Heerden, deputy director of the Louisiana State University Hurricane Center and a member of Team Louisiana, the group formed by the state to investigate the causes of the levee failures. His group, too, will be making proposals for upgrading protection for the region.

"We may get lucky," he said. "Nature may give us another 10 years before we get another Katrina, or maybe not. But we've got to seize the moment or we're going to lose coastal Louisiana."

Copyright 2006 The New York Times Company. Reprinted with permission.

In Review: Competing Plans to Repair New Orleans Flood Production

Testing Your Comprehension

1. As of the date of this article, how much of its planned repairs to New Orleans' flood protection system had the Army Corps of Engineers completed?
2. According to the article, many groups that have studied Hurricane Katrina are coming up with proposals of their own. What do they say are the benefits of these proposals?
3. What three things did the Bring New Orleans Back Commission propose to prevent storm surges from pushing into the city's drainage canals?
4. How much would these structures cost?
5. According to Lawrence Roth, of the American Society of Civil Engineers, what is the corps planning to do that might make the idea of jetties moot?
6. What three steps is the corps taking to restore flood protection to its pre-hurricane levels?
7. What is the primary concern of the people of New Orleans, to which the Bush administration has not committed?
8. According to Dan Hitchings, what aspect of the corps' program has put it slightly behind schedule?
9. What does the engineering society recommend to strengthen the levees?
10. What is the corps' reaction to this proposal?

Weighing the Issues

1. The Bring New Orleans Back Commission has proposed several "add-on" proposals that they believe will complement the Army Corps' current plans. Why do you think they feel that these add-ons are needed?
2. Beyond the actions described in the article, what other steps might be taken to protect New Orleans from flooding?
3. Describe three things that coastal communities can do to protect themselves against hurricane impacts that do not rely on new structures or technology.

4. What would be required to ensure the city's protection from Category 5 storms, "the toughest that nature can dish out"?
5. According to Lawrence Roth, "There is, very much, a tension between things that can be done quickly versus those that might take a little longer." What are the pros and cons of each?

More to the Story

To the fullest extent possible given the resources available to you, review and compare the plans of the Army Corps and the Bring Back New Orleans Commission. What are the strongest elements of each plan? Are these elements realistic? Why or why not? Combining elements of the two plans if necessary, outline the most effective elements of a comprehensive plan to repair New Orleans flood protection.

Useful Websites

http://www.mvn.usace.army.mil/

http://www.bringneworleansback.org

http://www.mvn.usace.army.mil/pao/response/amaps.asp

Answers to *Testing Your Comprehension* Questions

Clogged Rockies Highway Divides Coloradans (p. 3)

1. Unbridled growth, local identity, civic autonomy, and an uneasy dependence on government.
2. "How much do we really want to improve I-70 and do we want to improve it so much that it changes the character of our communities?"
3. Property values, boosterism, and the restless American impulse to move on and create anew.
4. Towns closer to Denver tend to favor mass transit, while those farther away favor highway expansion as a way to boost their economies.
5. Fifty percent.
6. A demographic transformation as members of the post-World War II generation reached their peak earning years, stock and real estate markets boomed, and changes in tax law made buying real estate more attractive.
7. $2 billion in recreation revenue and $136 million in state and local tax revenues.
8. Communities like Idaho Springs, confined in a narrow canyon, would be destroyed by asphalt, noise, and dust.
9. They worry that a mass transit line could turn resort communities into bedroom communities, full of commuters catching the train to Denver.
10. ". . . until you can get really high transit kinds of usage, the economics are not there."

Who Will Work the Farms? (p. 7)

1. Legal work channels for future immigrants and the current population of workers unlawfully in the country (carrots), and a much more effective bar against hiring illegal labor (sticks).
2. Guest workers cost more money, and you have to have illegal immigration under control.
3. It was intended to ensure that farmers could continue to get the cheaper foreign labor they wanted, but in a legally acceptable way.
4. Fewer than 25,000.
5. Seventy percent of 1.2 million, or 840,000.
6. Few employers were ever punished (also, the program is expensive).
7. It sets a floor for pay, and farmers must house the workers, pay workers' compensation, and pay all recruitment costs.
8. Legal workers are more dependable than illegal ones, and it relieves them of fears that their harvest may be in jeopardy if authorities conduct raids.
9. 1. To be entitled to use guest workers, they must certify that they cannot find American workers for the job. 2. They often find themselves being sued over working conditions.
10. In 2000, over 1,000 farmers employed 10,000 guest workers. In 2006, about 500 farmers employ about 5,000 workers.

Biotech's Sparse Harvest (p. 11)

1. By making it easier for them to control weeds and insects.
2. Public resistance to genetically modified foods, technical difficulties, legal and business obstacles, and the ability to develop improved foods without genetic engineering.
3. 1. Baby food companies were not interested because they are avoiding biotech crops. 2. Food companies feared lawsuits if some consumers developed allergic reactions to a product labeled as nonallergenic.
4. That the European Union has violated rules by halting approvals of new biotech crops.
5. In the sense that they have the gene that allows them to grow even when sprayed with Roundup.
6. 1. Soybeans high in omega-3 fatty acids. 2. Better tasting soy for use in products like protein bars. 3. Golden rice. 4. Crops able to withstand drought.
7. 1. Patent rights held by the big companies. 2. The cost of taking a biotech crop through regulatory review. 3. Technical issues. 4. Enhanced crops must meet the demands of farmers for high yields and of food companies for good taste and handling properties.
8. These crops do not face regulatory scrutiny.
9. Consumer benefits
10. Big food companies.

Deals Turn Swaths of Timber Company Land Into Development-Free Area (p. 15)

1. International Paper will receive $300 million for 217,000 acres in 10 states around the Southeast.
2. It will ideally revert to the cypress and longleaf pine forest that once covered the sandy flatlands.
3. Barely 2 percent.
4. 1. Based on market components, the forestlands are worth more to other people than they are to the company. 2. The deals offer International Paper the right to a supply of timber.

5. The amount of government and nonprofit money available is dwarfed by the amounts that can be offered by developers.
6. $140 million annually, down from $500 million in 2001.
7. A $32 million bond issue.
8. Their ecological value, including the presence of endangered species, the stock of hardwoods and softwoods, and the proximity of other protected areas.
9. 1. Available financing. 2. The company's preferences.
10. "They are the only folks who own very large pieces of land."

Americans Are Cautiously Open to Gas Tax Rise, Poll Shows (p. 17)

1. Americans are overwhelmingly opposed to a higher federal gasoline tax, but a significant number would go along with an increase if it reduced global warming or made the United States less dependent on foreign oil.
2. If it brought measurable results.
3. The development of alternative energy to reduce dependence on foreign oil.
4. 85%; 55%; 59%.
5. 1. Research for fuel cells. 2. Research for alternative fuel sources.
6. It might find its way into what they considered the wrong hands, just as the current tax supports highway maintenance and construction.
7. By lowering income taxes in a way that would "make most middle and lower income people better off."
8. By 6 to 8 percent "over the long run."
9. It would undermine pressure on people to buy fuel-efficient cars and move closer to their work, reducing their commute.
10. Funnel some of the additional tax revenue to manufacturers "as an incentive to offer more efficient vehicles, like hybrid cars."

The New Face Of an Oil Giant *Exxon Mobil Style Shifts a Bit* (p. 22)

1. He has gone out of his way to soften Exxon's public stance on climate change.
2. There is "still significant uncertainty around all of the factors that affect climate change."
3. Exxon's business is about increasing oil and gas supplies to consumers, not chasing alternatives that offer little prospect of replacing the fossil fuels that he views as the only realistic way to meet the world's growing demand for energy.
4. They plan to invest billions of dollars in the next decade to develop renewable energy sources like wind and solar power.
5. 31 percent as compared to 20 percent.
6. Easy-to-find oil has mostly been found, opportunities for new resources are scarcer, competition is rising, and governments are tightening the screws on international oil companies.
7. Gaining access to the world's fourth-largest oil field in the United Arab Emirates, and prevailing in a five-year-old dispute over the development of Indonesia's largest untapped oil reserves.
8. In the early 1990's, Exxon approached the Qatari government with an offer to serve as a joint partner. Today, Exxon is the largest foreign investor in Qatar and the nation is on track to become the world's leading liquefied natural gas producer.
9. Increase oil and natural gas production to five million barrels a day by 2010 and lift daily capacity by a total of two million barrels by 2030.
10. They made a 10-year, $100 million contribution to the Global Climate and Energy Project at Stanford, which focuses on long-term technological research.

Corn Farmers Smile as Ethanol Prices Rise but Experts on Foods Supplies Worry (p. 24)

1. Demand for ethanol.
2. High oil prices are dragging corn prices up with them, as the value of ethanol is pushed up by the value of the fuel it replaces.
3. More people will go hungry.
4. 1. To feed farm animals. 2. To make corn syrup for processed foods.
5. 1. Reduction of the need for oil. 2. Slowing of climate change.
6. It will cause instability in both energy and food prices.
7. It would provide new opportunities for small farmers and finance the development of roads and other infrastructure in poor rural areas.
8. Farmers might plan more corn and less wheat and cotton.
9. The U.S. is paying farmers not to grow crops on 35 million acres, and much of that land could come back into production.
10. More demand for soy oil as a diesel substitute would force production of meal, pushing down its price, and making cattle feed cheaper.

A Back-Fence Dispute Crosses an International Border (p. 28)

1. To protest the construction of a pair of paper mills there that they say will pollute the river that forms the frontier between the countries.
2. That they will damage agriculture and kill off tourism, all for the benefit of a couple of foreign companies.
3. $1.9 billion and 1.5 million tons each year.
4. 1. They accuse Uruguay of violating a treaty that governs use of the river. 2. They are irritated that their own president has not been more energetic in opposing the projects.
5. Meet the demanding environmental standards of the European Union and employ technology that reduces pollution to a minimum.
6. They charge that the promise of free movement in the founding charter of South America's Mercosur customs union is suffering.
7. The town has been "economically dead" for more than 20 years, and citizens welcome the 2,000 jobs and the revival of business activity the mills will bring.
8. Their country was created 180 years ago as a buffer between Brazil and Argentina, and throughout their history they have often complained of being bullied and scorned by their much larger neighbors.
9. Expressed interest in negotiating a free trade agreement with the United States.
10. Taking the case to the World Court in The Hague.

Recycled Inspection Reports and Other Water System Irregularities Stir Concerns Upstate (p. 30)

1. Court supervision over the watershed will end in August 2006.
2. Local officials may stop cooperating with the city, potentially costing New York billions of dollars.
3. He will urge the federal Environmental Protection Agency not to renew a crucial permit next year. Without that permit, the city could be forced to build a water filtration plant costing more than $8 billion.
4. The way the department takes care of its 22 dams.
5. That because of neglected maintenance, valves that might have been used to lower the reservoir's water level were buried under 15 feet of silt and could not be opened.
6. Seventy percent of the weekly inspection reports for dams on two reservoirs were actually photocopies of earlier reports, with only the date and weather conditions changed.
7. That what occurs now is the legacy of the department's predecessor, an autonomous agency created in 1905—long before the advent of environmental laws—to build the upstate reservoirs and pipelines. Conveying clean water to New York City was the priority, not complying with detailed federal regulations.
8. 1. Possessing about $10,000 in city property, including generators, snow blowers, and computers. 2. Falsifying water quality records at what is know as the Catskill Lower Effluent in Westchester County.
9. He shifted it to the city, and the workers who operate 14 huge wastewater plants.
10. That the violations reflect "a failure to grasp that compliance is everyone's job."

Advisories on Fish and the Pitfalls of Good Intent (p. 34)

1. Scientists who think it is important to eat tuna and farmed salmon for their omega-3 fatty acids, despite the contaminants they contain, versus those who think consumers should consider contaminants when deciding which fish to eat.
2. Methylmercury, PCB's, and dioxin.
3. Mercury may contribute to cardiovascular disease, neurological problems and immune system problems.
4. Through the United States Tuna Foundation and the Center for Consumer Freedom.
5. That those groups should eat no more than six ounces of albacore tuna a week, but could safely eat up to 12 ounces per week of low-mercury fish like shrimp, canned light tuna, salmon, pollock and catfish. They also continued to recommend that those groups limit their intake of shark, swordfish, king mackerel and tile fish.
6. That they should choose fresh, frozen, or canned wild salmon, which are comparatively uncontaminated with PCB's and dioxin.
7. They are concerned that such warnings confuse people and may make them reduce their consumption of all tuna and salmon.
8. They dropped 9.8 percent.
9. United States Tuna Foundation, National Food Processors Association Research Foundation, a trade association now known as the Food Products Association, and the Fisheries Scholarship Fund, part of the National Fisheries Institute, a seafood industry trade association.
10. If pregnant women ate low-mercury fish without decreasing total fish consumption there would be public health benefits, but if all adults ate less fish there would be a large negative impact on cardiovascular health.

Out of Old Mines' Muck Rises New Reclamation Model for West (p. 38)

1. It created a barrier that isolated the trout from competitors that could not fight their way through the muck.
2. It could make it possible for the competitors to move up.
3. Breckenridge and Summit County bought 1,840 acres of the 8,000-acre Horseshoe gold field from a company that had obtained the assets when the mines went bankrupt.
4. It's the first time that the E.P.A. entered into an agreement with a purchaser of a site like this who voluntarily agreed to do a cleanup.
5. They will spend $2 million for environmental cleanup, including a $1.2 million water-treatment plant that would have to be operated in perpetuity at a cost of about $90,000 a year, on top of the $8.2 million purchase price.
6. Because the risks are substantial. Just digging a hole, if it uncovers or disturbs environmental problems, could expose the new owners to liabilities.
7. That the property's former owners would begin selling off less damaged parcels to pay for the cleanup and water treatment that the E.P.A. wanted.
8. Town: wants hiking and biking. County: wants some motorized use.
9. 1. It is a narrow monoculture. 2. Pine beetles could kill 80 percent of the trees.
10. The artificial waterfalls from the pipes under road crossings have provided adequate barriers to downstream interlopers, clean water or no.

New Uses for Glut of Small Logs From Thinning of Forests (p. 40)

1. Intense forest fires burned more than 350,000 acres of forest.
2. To reduce the wildfire threat to towns surrounded by publicly owned forests.
3. A glut of logs smaller than eight inches in diameter.
4. Its high cost, low value, and a lot of pieces to handle, which takes time and effort.
5. $90 a ton for house logs, less than half that for the smaller trees.
6. 1. Heating schools and hospitals. 2. Construction materials, including particleboard, flooring, and laminated beams.
7. Because there are so few big trees left.
8. To give grants up to $400,000 for schools and other public buildings to build furnaces that burn biomass.
9. $125,000 with diesel, $28,000 with biomass.
10. He wants "to make sure it's done well and done right, so we don't create more demand than the land can sustain."

Canada to Shield 5 Million Forest Acres (p. 44)

1. It is a reminder to future generations that the world must be kept pristine.
2. To keep a wilderness of close to five million acres off limits to loggers, and to undertake selective logging in 10 million other acres.
3. Grizzly bears, unique subspecies of goshawks, coastal wolves, Sitka blacktail deer and mountain goats.
4. That "it's about approaching business with a conservation motive up front, instead of an industrial approach to the forest."
5. They will be obliged to cut selectively; away from critical watersheds, bear dens, and fish spawing grounds.
6. Areas with cultural significance like ancient cemeteries, or those with medicinal herbs and cedars big enough to make totem poles, canoes, or long houses.
7. Ecotourism lodges, shellfish aquaculture and other environmentally sustainable economic activities.
8. One quarter.
9. An estimated 20 percent.
10. Postcards, demonstrations, and the Home Depot's announcement to phase out sales of wood from the forest.

Competing Plans to Repair New Orleans Flood Protection (p. 46)

1. Sixteen percent.
2. That they could be cheaper, faster, or stronger than the corps' plans
3. 1. A series of jetties to stand in front of the three canals. 2. A network of dams that would block or slow storm surges and the opening between the Inner Harbor Navigational Canal and Lake Pontchartrain. 3. Long-term flood-control structures that would block or slow surges at the two passes between Lake Pontchartrain and the Gulf of Mexico.
4. $100 million to $170 million.
5. Closing off the canals and putting in new pumping stations at the lake.
6. 1. Soliciting bids for building temporary closures and pumps at the mouths of the city's drainage canals.

2. Rebuilding stretches of levee. 3. Studying how to strengthen the nurrican protection system.

7. Protection from Category 5 storms.
8. The "gear-up time" to get contractors in place and to make materials available is sizable.
9. "Armoring" the dry side of the levees so they are not eroded away from underneath if water spills over the top.
10. That the armoring process will have to be approved by Congress.

Answers to *Interpreting Graphs and Data* Questions

Clogged Rockies Highway Divides Coloradans (p. 4)

1. Eagle
2. Clear Creek
3. Eagle
4. It should increase the rate of growth of each county.

Americans Are Cautiously Open to Gas Tax Rise (p. 18)

1. Reduced dependence on foreign oil.
2. 4%

The New Face of an Oil Giant (p. 22)

1. EM: +47.4%. BP: +12.0%. C: 30.4%.
2. EM: $3.5 billion. BP: $250 billion. C: $225 million.
3. EM: $138.8 billion. BP: +107.1 billion. C. +79.1 billion.
4. EM: $35 billion. BP: $20 million. C: $10 million.
5. EM: $18.4 billion. BP: $12.5 billion. C: $6.3 billion.
6. Chevron.